Hitting the Wall

A Vision of a Secure Energy Future

Hitting the Wall: A Vision of a Secure Energy Future
Richard Caputo

ISBN: 978-3-031-79422-3   paperback

ISBN: 978-3-031-79423-0   ebook

DOI: 10.1007/978-3-031-79423-0

A Publication in the Springer series

SYNTHESIS LECTURES ON ENERGY AND THE ENVIRONMENT: TECHNOLOGY, SCIENCE, AND SOCIETY #3

Lecture #3

Series Editor: Frank Kreith, Professor Emeritus, University of Colorado

Series ISSN
ISSN 1940-851X   print
ISSN 1942-4361   electronic

Hitting the Wall

A Vision of a Secure Energy Future

Richard Caputo

SYNTHESIS LECTURES ON ENERGY AND THE ENVIRONMENT: TECHNOLOGY, SCIENCE, AND SOCIETY #3

ABSTRACT

Hitting the Wall examines the combination of two intractable energy problems of our age—the peaking of global oil production and the overloading of the atmosphere with greenhouse gases. Both emerge from the overconsumption of fossil fuels and solving one problem helps solve the other. The misinformation campaign about climate change is discussed as is the role that noncarbon energy solutions can play. There are nine major components in the proposed noncarbon strategy including energy efficiency and renewable energy. Economics and realistic restraints are considered and the total carbon reduction by 2030 is evaluated, and the results show that this strategy will reduce the carbon emission in the United States to be on track to an 80% reduction in 2050. The prospects for "clean" coal and "acceptable" nuclear are considered, and there is some hope that they would be used in an interim role. Although there are significant technical challenges to assembling these new energy systems, the primary difficulty lies in the political arena. A multigenerational strategy is needed to guide our actions over the next century. Garnering long-term multiadministration coherent policies to put the elements of any proposed strategy in place, is a relatively rare occurrence in the United States. More common is the reversal of one policy by the next administration with counterproductive results. A framework for politically stable action is developed using the framework of "energy tribes" where all the disparate voices in the energy debate are included and considered in a "messy process." This book provides hope that our descendants in the next century will live in a world that would be familiar to us. This can only be achieved if the United States plays an active leadership role in maintaining climatic balance.

KEYWORDS

peak oil, climate change, global warming, energy efficiency, renewable energy, solar, geothermal, wind, biomass, bioliquids, photovoltaics, distributed solar, national grid, pluggable hybrid electric vehicle, energy policy, concentration solar, energy tribes, acceptable nuclear, clean coal, policy, U.S. leadership, transportation efficiency, building efficiency, industrial efficiency

Acknowledgments

Without Frank Kreith's support and encouragement, not a word would have been written. Joel Claypool continued this support until there was a final draft. This book would not be technically possible with such a depth of comprehension without the permission of the American Solar Energy Society via Brad Collins, executive director, to liberally use the material in their vital report, "Tackling Climate Change in the U.S." [Charles F. Kutscher (ed.), ASES, January 2007], This key report is the brainchild of Charles Kutscher, who used the National ASES Conference in 2006 in Denver to generate the nine key papers that are at the core of noncarbon approaches part of this report. These conference papers were turned into the vetted Tackling Climate Change report. The individuals who are responsible for these nine definitive sections are Joel N. Swisher, who developed the overall energy efficiency section; Marilyn A. Brown, Therese K. Stovall, and Patrick J. Hughes, who generated the building energy section; Peter Lilienthal and Howard Brown, for considering the plug-in hybrid electric vehicle; Mark S. Mehos and David W. Kearney, who displayed the concentrating solar power situation; Paul Denholm, Robert M. Margolis, and Ken Zweibel, who contributed the photovoltaic power section; Michael Milligan, for developing the wind power story; Ralph P. Overend and Anelia Milbrandt, for their insights into the biomass opportunities; John J. Sheehan, who laid out the likely contribution from biofuels; and Martin Vorum and Jefferson W. Tester, who exposed the possibilities of geothermal power. Besides designing the approach to properly consider these noncarbon solutions to the climate change problem and engaging these talented researchers to contribute in their areas of expertise, Chuck Kutscher pulled together the results and placed them in the framework of the climate change problem. I owe a deep debt to this team for developing the energy efficiency and six renewable opportunities in a comprehensive and consistent manner.

I am in debt to several other researchers who developed substantial analysis of key areas that I depended on heavily. Lisbeth Gronlund, David Lochbaum, and Edwin Lyman gave me clear insight into the nuclear industry and what needs to be done before seriously considering using nuclear power as part of our energy future. I relied on Kenneth Deffeyes and C. J. Campbell who gave me insights into the current global oil situation. I depended on Jeff Goodell for his work in the coal industry. Carl Baker was generous with his insights on long-distance transmission issues. The gained insights into the different "energy tribes" and how to base policy on these concepts were provided by Marco Verweig and Michael Thompson.

I gained much from the review of early drafts by Kenneth Deffeyes, Jeff Severinghaus, Peter Lilienthal, Mark Mehos, Anelia Milbrandt, Lisbeth Gronlund, John Sheehan, Anelia Milbrandt, Marco Verweig, Carl Barker, Simon Shackley, and Celia Orona. Finally, Carol Jacklin provided daily support in a host of tangible and not so tangible ways. I am deeply indebted to her.

All of the brilliant insights in this book are to be credited to all those cited above. All the inappropriate interpretations and errors of judgment must be laid at my feet.

Dedication

To Carol, who found the space to let me have my space.

Contents

CHAPTER 1

Introduction

What could have a more severe impact on Americans than any terrorist's action? Two greater threats to the United States are examined to scope their magnitude and identify approaches aimed at reducing their impact. Challenging technical, economic, and political dimensions remain.

The current peaking in global oil production coupled with strong demand will have a dramatic impact on the U.S. economy given its heavy reliance on oil. The obvious result will be sharply rising oil prices with massive economic impacts. Because of the relative inelasticity of oil consumption in the United States, the only short-term mechanism left to match stagnating supply with increased demand is a sharp economic downturn. This is characterized as hitting the oil price wall.

Simultaneous with this pending economic disaster is the overloading of the atmosphere with excess carbon and other greenhouse gases. Unless stringent measures are taken, human activities will drive carbon dioxide (CO_2) levels to exceed 500 ppm in the next few decades and cause the temperature to rise beyond anything know in the recent geological past. This human-caused "consumption" of the atmosphere's ability to absorb CO_2 over the past 200 years comes at a high temperature point of the interglacial period and will drive the planet's climate behavior into unknown territory. A sober estimate of the economic consequences of this climate-altering future will cost 5% to 20% of global GNP (Stern Report). This phenomenon is called hitting the CO_2 wall. Taken together, these significant events can simply be called hitting the wall.

Dealing with these events is probably the most difficult challenge our species has faced since our inception 200,000 years ago. The difficulties are technical and economic, but most of all, political. Although a significant challenge, technical opportunities do exist to move beyond the use of carbon as a fuel. These extend from a host of energy efficiency prospects that give the same service at a fraction of the energy now used to a wide range of renewable energy opportunities. There may also be occasions to use "acceptable" nuclear power and even "clean" coal as interim energy sources. However, the real difficulty lies in the political realm.

At the level of individuals, there are such grossly different views on these global issues that it almost defies understanding. The very same events (peaking of global oil production and out-of-control carbon emissions) elicit a response that ranges from "these are not problems, they are opportunities" to "these are the most significant problems ever faced by humankind."

On the national level, you have the most significant historic polluter (United States) as the only county that has not adopted the climate change treaty (Kyoto) and the two emerging significant polluters (China and India) refusing to take action until the United States does. Beyond this current stalemate, there is the intractable-appearing conflict between the people and the planet. We have come to rely on "cheap" and abundant carbon as a fuel to power us from a preindustrial existence to the giddy heights of our highly industrialized societies. The use of carbon is almost solely responsible for raising the standard of living for almost all of the industrialized world. However, this is less than 15% of the world's population. This small number of people has used up the common atmosphere and have left it overburdened with CO_2 and other greenhouse gases. The other 85% have no hope of also using fossil fuels to raise themselves out of poverty without further destruction of the planet. To limit the damage to the planet, we all have to reduce our carbon emissions by about 50% by the year 2050. How can this be arranged while giving 85% of the world's population a chance at a decent future?

The difficulty with this problem is marked by the stark differences in perceptions at the individual level of what is a problem and what is a solution, and the inequities in the consumption of the planet's atmosphere among nations that have yet to demonstrate much skill at arriving at equitable solutions.

This book examines the magnitude of the oil peak and carbon emission problems and examines technical and economic solutions. Policy opportunities are laid out and examined in light of the different individual perceptions, and the minimum requirements for a satisfactory political solution are noted. Policies are suggested to address the massive inequities that exist and yet limit carbon emissions to avoid the worse of the expected impacts.

· · · · ·

CHAPTER 2

The End of Cheap Oil

2.1 GLOBAL OIL PEAK PRODUCTION—WHEN?

The United States is still a major oil producer (third largest in the world) and has supplied its own oil from the start of production at the Drake well in Pennsylvania in 1859 until 1946 when the United States started importing more oil than it exported. In 1970, our domestic production peaked after 111 years of production. Because our use of oil was still growing, the shortfall was made up by importing 40% of our oil use. A politically induced supply slowdown by the Organization of the Petroleum Exporting Countries (OPEC) in 1973 (Yom Kippur War oil embargo) caused significant price increase and exposed the vulnerability of the United States to importing oil from a politically unstable region. After several false starts to achieve energy independence beginning with Richard Nixon, we now import 60% of our oil.

OPEC (the Middle East-based oil cartel) normally attempted to keep oil price in the $22 to $28/bbl range [1] to reap large profits since the cost to actually produce Middle East oil was only a few dollars per barrel. This price level was also designed to block the introduction of oil substitutes. Although the average world price of oil was in this range in 2006 dollars, it varied wildly from $10 to $70 per barrel (see Figure 2.1).

There was a reversal of policy starting with the Reagan administration, and the United States was supportive of this OPEC attempt to moderate oil pricing and abandoned the attempts to achieve energy independence. The United States was indifferent to its longer-term interests and cooperated with oil priced too low or too variable for alternatives to enter the market. The United States put off the day when we would have to develop our own energy policy to better meet our long-term needs.

In the United States, oil production has steadily dropped since the 1970 peak even with new oil fields being found and developed such as the large Alaskan fields and new technology being used. The oil extraction rate followed a "bell-shaped" curve around the 1970 peak. This oil production peak was predicted in 1956 by M. King Hubbert (a geologist at the Shell research laboratory in Houston who taught at Stanford and UC Berkeley, worked at the U.S. Geological Survey, and was a member of the National Academy of Sciences). At the time, there was controversy about

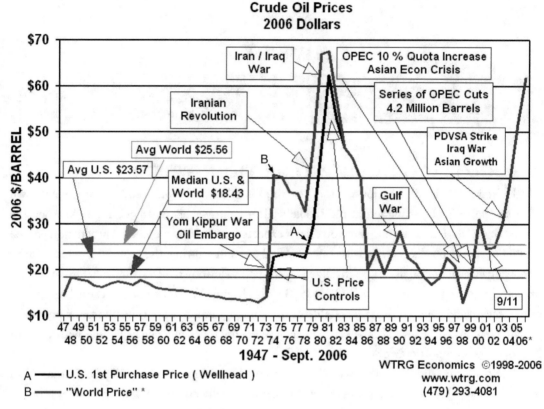

FIGURE 2.1: Crude oil prices: 2006 dollars.

his prediction, but 14 years later, he was proven right. Basically, he based his prediction on the annual rate of oil production compared to the cumulative oil taken out of the ground. As the ratio of these two quantities fell due primarily to fewer and fewer new discoveries, it became clear that production would soon follow and start falling.

Using the same geologic-based approach, credible analysis now predicts that global oil production is peaking in this decade [2–4]. Figure 2.2 shows the drop in new discoveries and the relationship to production.

This peak of production in this decade signifies that we are halfway through the total recoverable oil on the planet as illustrated in Figure 2.3. Thus, in 150 years, we have used half of the global oil reserve that took hundreds of millions of years to lay away for us [5]. Recent data indicate that global oil production has been constant since 2005 and possibly is the early indicator that we are

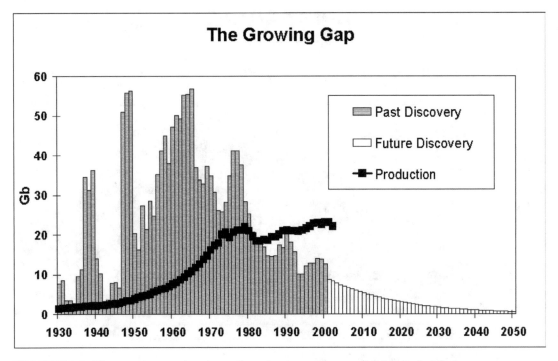

FIGURE 2.2: The growing gap between oil production and new oil fields finds [5].

at the peak. There is some disagreement about the this mid-way point of global oil consumption, and the largest estimate of remaining recoverable oil is made by the U.S. Geological Survey that predict a peak in 2030. There is a well-deserved skepticism of this view. No matter who you think is more credible, the difference in the peak year prediction ranges from this decade to as far away as 2030 . Either way, they both are saying the same thing because the transition away from oil will take decades. The only difference is that one group of experts say you should have started 20 years ago to do something about this peaking of global oil production this decade, and the other agency is saying that you should start now in anticipation of the 2030 estimated peak.

Currently, there is a 3% annual growth in oil demand in the market. This growth rate is due to (1) the United States, the dominant oil consumer, renewing high consumption after the 1980s due to rapid growth of large SUVs and light trucks for personal use and increases in miles driven; (2) developing countries such as China and India with large economic growth and the resulting increased use of cars; and (3) the global economy being fairly robust up to 2008. Peaking global oil production means that production from all sources will be flat for a few years and then start to

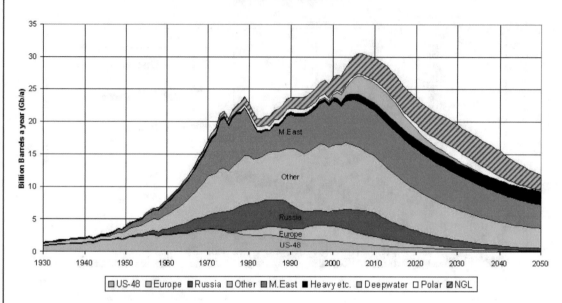

FIGURE 2.3: Oil and gas liquids: 2004 scenario.

decrease about 2% per year. The current 3% growth cannot be managed in a world with production dropping by 2%. Something has to give!

2.2   ALTERNATIVE SCHOOLS OF THOUGHT TO ALTERNATIVES

Because oil is the mainstay of our ground and air transportation industries, it is quite a challenge to come up with a timely transition to alternatives. There are three schools of thought about the policy course that makes the most sense: the neocon use of military force to gain control of Middle East oil and sidestep OPEC; the "market will know what to do" approach advocated by liaise-fair economists and conservative politicians; and the "switch to nonfossil energy sources" approach. Each approach seems to have some merit, but only one can take us to a future without excessive risk of disruption and with a measure of self-control.

2.2.1   Neocon Insight

The neocons played a peripheral role in the first Bush administration but came to power in the second Bush administration. As initially oulined by Paul Wolfowitz and "Scooter" Libby, they took what can only be called a muscular (bull in the china shop) approach to what they saw as the major threat to U.S. international dominance for the next 50 to 100 years. (Patrick Tyler, "U.S. Plans Calls for Insuring No Rivals Develop", New York Times, March 8, 1992) As the only superpower, the United States has total military dominance based on its spending as much on armaments than all the rest of the world combined (Stockholm International Peace Research Institute at http://www. sipri.org/). The only real threat to the long-lived dominance of the United States was the high cost and instability of the price of oil. This periodic volatility would be disruptive, cause our economic growth to falter, and show that we were not dominating the world. The cause of this fly in the oil is OPEC, which started in 1960 when Venezuela talked four Middle Eastern countries into forming this cartel to control oil price. The key to OPEC's success is Saudi Arabia, which has the lowest production cost and greatest reserves of oil.

Currently, the Saudis are considered a weak link in the OPEC armor with the country's finances in shambles with the fabled oil fortune from the 1970s depleted by the opulent lifestyles of about 15,000 royal prices along with an extensive welfare state with expanding population. In addition, there are the hundreds of millions per year that the royal family is spending to placate Islamic fundamentalists. This funding is funneled primarily into schools that preach a doctrine that is relatively hostile to the lifestyle of the royal family that is viewed as corrupt. These schools seemed to be exporting radical Islam.

The first war ever fought exclusively over oil was when dozens of countries led by the United States went to the Middle East to expel Iraq from Kuwait in 1990. The thought of having a country such as Iraq control a more substantial part of world oil with potential motives beyond greed solidified support for the U.S. led repulse of Iraq. The first Bush administration showed some insight when it refused to march up to Baghdad to take over Iraq at the conclusion of the Kuwait war. The refusal was not based on any thought of military difficulty but on the prospect of losing the broad base of support for the war that legitimatized and paid for the effort, and the instability that removing Saddam would bring to the Middle East.

The neocon-controlled second Bush administration had a vision about how to enable the United States to dominate the world for decades. In fact, in the very first Bush international cabinet meeting 10 days into the new administration at the end of January 2001, the neocons had already decided without cabinet discussion that the major problem in the international sphere was Iraq. The only question was how to manage regime change and gain control of Iraq's oil [6].

The reason this was important was because the neocons estimated that the 3.5 million barrels a day preinvasion Iraq oil production could be raised to 7 million barrels a day by 2010. This would cost $5 billion initially just to resume prewar production levels. An additional 40 billion dollars would be needed to push production up, and this money could only come from Western oil companies. This investment would only occur if the outside oil companies received guarantees they would have a healthy share of revenues and that the market, not OPEC, would determine future production and price.

With U.S. occupation and control of Iraq, the neocons thought it was likely that Iraq could be convinced to ignore its OPEC quota and start producing at maximum capacity. All this new oil would effectively end OPEC's ability to manipulate price. As the price of oil dropped, Saudi Arabia would be caught between its high revenue needs and low oil prices. The Saudis, the neocons figured, would also be forced to open their oil fields to Western oil companies as would other OPEC countries. With oil at a free market value of between $10 and $15/bbl, Americans would have long-term economic growth unhampered by the threat of high and volatile prices. Combined with military hegemony, this would allow the United States to prevail for a half century or more. Quite a radical vision [7]!

Well, that was the reason we manufactured the rationale for the preemptive Iraq war out of "whole cloth." The cloth was woven by a special unit created by Donald Rumsfeld called the Office of Special Plans, which he started in September 2002. This Pentagon unit was led by Douglas Feith and dealt with "reshaping" intelligence on Iraq. Because the normal intelligence reports were not producing the "right" answers on Iraq, this unit picked over the flotsam and jetsam of the pile of intelligence material to put together the information desired by Rumsfeld. This misuse of intelligence was finally officially reported by the Inspector General of the Pentagon on February 9, 2007.

We now have the answer to the question asked by the long-time member of the White House press corps, journalist Helen Thomas, who in early 2006 was called upon directly by President Bush for the first time in 3 years. Thomas asked Bush about Iraq, ". . . your decision to invade Iraq has caused the deaths of thousands of Americans and Iraqis . . . . Every reason given, publicly at least, has turned out not to be true. My question is: why did you really want to go to war? . . . what was your real reason? You have said it wasn't oil—quest for oil, it hasn't been Israel, or anything else. What was it?" What is clear now it that the publicly stated reasons such as Iraq harboring weapons of mass destruction based on nerve gases or nuclear weapons through rockets as means of delivery were erroneous as was the alleged connection between Iraq and the al-Qaeda terrorists.

Most Americans now know that this was a deeply flawed dream on the part of the neocons. Their actions to declare a preemptive war without abiding by civilized norms stated in the UN convention have morally bankrupted the United States and revealed us to be power-crazed militarists who are devoid of ethics. The neocons have lost the thing we prize most—our sense of collective decency that was squandered for a maniacal pipe dream that was totally unrealistic. Besides the dead Iraqis and Americans and continued instability in the world's richest oil region, we have lost and will lose a tidy sum of money with the out-of-pocket costs now close to $500 billion. The expected extended costs beyond the direct conflict will be about another $500 billion. So we are talking about close to 1 trillion dollars [8]! Although it is only money, it is interesting to see how misguided the neocons are in yet one more area when we note that former Deputy Defense Secretary, Paul Wolfowitz, had said Iraq could "really finance its own reconstruction."

2.2.2 The Market Will Signal

What about the second school of thought on what to do about the impending peaking of global oil production? Some economists and politicians expect that there will be a price signal as oil becomes more dear and that will trigger the investments in alternatives. There are some difficulties with this expectation that the market will foster a timely transition away from oil before hitting the wall. This strategy is pretty straight economics 101 and acceptable for college sophomores, but it makes a number of erroneous assumptions. The key one being that there is no perfect market functioning in international oil transactions. The OPEC cartel routinely manipulates oil price to be in their sweet spot (too low for alternatives and high enough for a good profit). Although the Saudi production cost is a few dollars a barrel, the price has been higher than $30/bbl since 2002 (Energy Information Agency).

For a price signal to work, a free market exchange is needed. This is where the cheapest production price oil is produced first and sold at a profit. Then, the oil that is more and more expensive to produce comes to the market and there is a gradually increasing price as cheaper sources are exhausted. This rising price would start sending a larger and larger signal to bring in alternatives. Also, there would be little opportunity for extreme variability as what exists now with OPEC manipulation. Large price variability is a daunting barrier to large investments in alternatives.

A second difficulty with a timely transition away from oil due to a gradual price signal is oil's price inelasticity as shown in the 1970s. After each of the six major oil price spikes since the Second World War, global economic activity had begun to fall within 6 months: typically, every $5 increase in oil prices brought a 0.5% decline in economic growth. Even after this oil price spike subsisted, the economic recovery gained only a tenth of what they had lost in the spike. Cumulatively, price spikes had cost the economy 15% in growth and more than $1.2 trillion in direct losses [9]. So there is a large economic price paid for an oil price spike, and it is difficult in the short term to reduce oil use gracefully to avoid this economic price.

The "market will decide" school of thought is not much help in dealing with the global peaking of oil production. This certainly is a situation when government regulation is needed to affect a desirable outcome.

The only effective short-term mechanism to use less oil is a strong economic downturn. Other mechanisms take some time. When the oil price shock of the early 1970s hit, it took the United States and Western Europe about a decade to start significantly reduce oil consumption. The United States used regulation (mandatory vehicle fleet miles per gallon standards; Corporate Average Fuel Economy) and Europe used a gasoline tax of about $3.50 per gallon. Although these two policy options are quite different, they both achieved the same result, and by the mid-1980s, the average miles per gallon of vehicles doubled in both the United States and Western Europe. These effective actions took about a decade to produce results. This is also about the length of time for a large new oil field to start producing. Improved car efficiency standards are "forever," and the new oil field may pump for a decade or two.

When the peak of global oil production is reached, the economic dislocation will not be temporary. There is no higher level of oil production to go to get another reprieve from the basic problem. Global is as big as it gets on the planet. This time, without effective measures, there would be major and long-term economic dislocation that would be either a severe long-term recession or global depression. Some are calling it the Second Great Depression [10]. I call it hitting the "oil price wall."

Out of this economic chaos will come a call for a "quick fix" that will likely be a draconian solution that is not in our long-term interests. Most people at that point would not care about our long-term interests. They would want to straighten out the turned over economic wagon and clutch at straws. Without changes in national leadership and based on a linear extension of the U.S. energy bill of 2005, draconian solutions will likely abandon environmental interests as well as the current democratic public involvement process in decision making. An "energy czar" will make decisions in the "back room." The form of the draconian solutions will most likely be:

- More oil drilling without restraint in preserves and in offshore areas that are now restricted;
- Attempts to use coal to generate transportation liquids with massive increase in carbon emissions;
- More natural gas drilling especially coal gas without environment check;
- More coal mining and combustion without pollution controls resulting in more air pollution deaths, acid rain, and mercury poisoning;
- Restart of the American nuclear industry without design changes that would avoid core meltdowns with the potential for catastrophic accidents and without the long-term solution of nuclear wastes;
- No public input on siting or on any other feature of these energy choices, and no environmental impact statements to consider alternatives.

In sum, this approach represents a significant disruption of the environment and a destruction of the democratic public process in these decisions—a draconian solution indeed.

Only two items in the list has anything to do directly with oil production, and even this option does not provide relief in the near term. A new oil field takes about a decade to develop and reach significant production. With other domestic oil production continuing to drop on the down side of the bell-shaped curve, this new high-impact oil production would not solve the production problem. Use of coal to make transport liquids is difficult, expensive, and would increase pollution especially carbon emissions.

For example, we have referred to the lower 48 oil production peaking in 1970. What happened when you include all 50 states? When Alaska, with its considerable oil production, was included in total U.S. oil production, there was no second and later peak of U.S. oil. The lower 48 oil had continued to decline and offset the new Alaskan oil flow. Even if you used the estimates of Artic National Wildlife Refuge that are optimistically high, at current worldwide oil use of 25 billion barrels per year, the 5 billion barrels from the Artic National Wildlife Refuge would postpone the world decline for 2 to 3 months.

Thus, allowing oil production in restricted areas will not cause U.S. oil production to increase. The continued tail-off in existing oil fields will continue to drive down overall U.S. oil production. Also, this unrestricted oil mining is not a solution. It would flatten out the downside of the bell-shaped curve a bit. You would still have to do something more significant to solve the problem.

The other draconian options identified above could generate more energy, but they do not directly contribute to liquid fuels for transportation. However, they can contribute to increased electricity generation that can contribute to the transportation sector via direct use of electricity or much later, via the hydrogen energy carrier. Again, not a near-term option. The pluggable hybrid electric car is a nearer-term option. Using electricity from an energy system dominated by burning pulverized coal would not reduce pollution. The transition from petroleum- to electricity-driven cars would also take decades. Decades would be involved in finding a substitute for oil even with the draconian approach.

After reviewing the national energy policy act of 2005 in this light, it is a clear first step in the direction of the draconian. A frightened public will be demanding that something be done to avoid the economic instability and chaos resulting from the rapid increasing price of oil. Americans show an amazing characteristic to be manipulated by conservative forces using fear tactics as what the support for the Iraq war demonstrates most recently. When the oil price wall is hit in the next decade or so, that their will be a great temptation to put the final pieces of this draconian approach will be put in place.

2.2.3   Switch to Alternatives

That seems to leave the third and final school of thought about what to do about global oil production peaking. This switching to nonfossil domestic fuel energy sources is our only real choice. However, it is not just up to the United States. Even if the United States significantly changes how it does energy business and adopts the political measures to redirect our energy system to a long-term renewable approach, there still is the issue of other nations using too much oil. This alone can cause the massive oil price shock we are trying to avoid. To make this domestic policy work, there would have to be cooperation among all oil-using nations to implement similar measures.

Fortunately, many nations are giving these sorts of measures serious consideration because of global warming (Kyoto Agreement at http://unfccc.int/kyoto_protocol/items/2830.php). All of the recommended measures presented in the following chapters would make a strong contribution to reducing the rate of global warming. This recommended strategy has two duties—it helps avoid a draconian future, which is a mix of an environmental disaster and weakening of our democratic fabric and the associated oil price-induced world economic downturn, and it directly contributes in solving the intractable problem of global warming. This alternative strategy is developed in chapter 4, the carbonless energy options.

. . . .

CHAPTER 3

Carbon—Too Much of a Good Thing

Curiously, the peaking of global oil production dovetails into the climate change issue, the other planetary-sized issue occurring at this time. Both problems are directly linked with the excess use of carbon. It is time to move past the polite term *global warming* and speak directly of "climate change" or even better, "global heating." The rapid global climate changes now under way are directly linked to human activities. Thousands of scientists from 120 nations mobilized since 1988 into the Intergovernmental Panel on Climate Change (IPCC) who study climate change reached consensus about the role human activities play in causing climate change. They have produced this consensus-based analysis with increasing probability in Assessment Reports in 1990, 1995, 2001, and 2007. The latest and fourth report increased the confidence in human activities causing the current heating trend. Human causality probability went from the range of 60–90% based on the 2001 report, to 90% certainty at present. As the data firmed up over the past 6 years and the analytical tools improved, the basis for scientifically establishing the human causality increased to a level that there is no credible basis for doubting the smoking gun of human causality.

3.1 CLIMATE CHANGE PARALLELS TOBACCO INDUSTRY "SCIENCE"

The IPCC goes through great lengths to involve accredited nongovernmental organizations from all sides of the issue. Known contrarians such as Richard Lindzen and John Christy have been involved as report chapter lead authors and contributing authors. Other figures, such as Fred Singer and Vincent Gray, have been involved in the process as reviewers. There is some uncertainly over the timing and magnitude of the climatic impacts due to climate change, but there is no disagreement over the conclusion that human activities are driving the rapidity of climate change.

A massive study of all the peer-review journal articles on climate change over the last decade has shown that there is total agreement on the conclusion that current climate change trends are due to human activities [11]. There were no scientific articles that opposed this view when a significant sample was checked in all journal publications over a decade. There is no debate about the science. The scientific consensus is summed up by Dr. Robert Watson, then Chairman of the IPCC, who said in 2001, "The overwhelming majority of scientific experts, whist recognizing that scientific uncertainties exist, nonetheless believe that human-induced climate change is already occurring and that future change is inevitable."

However, there is a lively cottage industry in popular media on this topic. A small cadre of writers is actively introducing "doubt" about the scientific agreement on the human causes of climate change. It is worth taking a quick look at this force in society to try and understand why, in spite of the science, so many Americans will say, "climate change is part of the natural variability of things and what's happening now is not caused by human actions." This aspect of Climate Change media coverage was investigated in some depth by the Union of Concerned Scientists [12].

This effort to produce just this result is lead by the Global Climate Coalition (GCC) founded in 1989 by 46 corporations and trade associations representing all major elements of U.S. industry. The GCC presents itself as a "voice for business in the global warming debate." The group funded several flawed studies on the economics of the cost of mitigating climate change, which formed the basis of their 1997–1998 multimillion dollar advertising campaign against the Kyoto Protocol. The GCC began to unravel in 1997 when British Petroleum (BP) withdrew its membership. Since then, many other corporations have followed BP's lead, and this exodus reached a high point in the early months of 2000 when DaimlerChrysler, Texaco, and General Motors all announced their exodus from the GCC. Since these desertions, the GCC restructured and remains a powerful and well-funded force focused on obstructing meaningful efforts to mitigate climate change. Their main speaking point is that climate change is real, but it is too expensive to do anything about, and that the Kyoto Protocol is fundamentally flawed. Major funding is from corporate members (industries, trade associations, etc.) [12].

The George Marshall Institute is a conservative think tank that shifted its focus from Star Wars to climate change in the late 1980s. In 1989, the Marshall Institute released a report claiming that "cyclical variations in the intensity of the sun would offset any climate change associated with elevated greenhouse gases." Although refuted by the IPCC, the report was very influential in influencing the Bush Sr. administration's climate change policy. The Marshall Institute has since published numerous reports downplaying the severity of global climate change. Their primary message is to blame the Sun and that the Kyoto Protocol is fatally flawed. Affiliated individuals are Sallie Baliunas, an astrophysicist from Harvard, and Frederick Seitz.

The Oregon Institute of Science and Medicine (OISM) teamed up with the Marshall Institute to cosponsor a deceptive campaign known as the Petition Project, whose purpose was to undermine and discredit the scientific authority of the IPCC and to oppose the Kyoto Protocol. Early in the spring of 1998, thousands of scientists around the country received a mass mailing urging them to sign a petition calling on the government to reject the Kyoto Protocol. The petition was accompanied by other literature including an article formatted to mimic the journal of the National Academy of Sciences. Subsequent research revealed that the article had not been peer reviewed, nor published, nor even accepted for publication in that journal, and the Academy released a strong statement disclaiming any connection to this effort and reaffirming the reality of climate change. The Petition Project resurfaced in 2001. OISM's main talking point is that there is no scientific basis for claims about climate change, IPCC is a hoax, and Kyoto is flawed. Funding is from private sources. Affiliated individuals are Arthur B. Robinson, Sallie L. Baliunas, and Frederick Seitz.

The Science and Environmental Policy Project (SEPP) was founded in 1990 by widely publicized climate skeptic S. Fred Singer, with SEPP's stated purpose is to "document the relationship between scientific data and the development of federal environmental policy." SEPP has mounted a sizable media campaign of publishing articles, letters to the editor, and a large number of press releases to discredit issues of climate change, ozone depletion, and acid rain. Their main message is that climate change will not be bad for us anyway. Action on climate change is not warranted because of shaky science and flawed policy approaches. They receive their funding from conservative foundations including Bradley, Smith Richardson, and Forbes. SEPP has also been directly tied to ultraright-wing mogul Reverend Sung Myung Moon's Unification Church, including receipt of a year's free office space from a Moon-funded group and the participation of SEPP's director in church-sponsored conferences and on the board of a Moon-funded magazine. Individuals associated with SEPP are Fred Singer and Frederick Seitz [12].

The Greening Earth Society (GES) was founded on Earth Day 1998 by the Western Fuels Association (WFA) to promote the view that increasing levels of atmospheric carbon dioxide (CO_2) are good for humanity. GES and Western Fuels are essentially the same organization. Both used to be located at the same office suite in Arlington, VA. Until December 2000, Fred Palmer chaired both institutions, and the GES is now chaired by Bob Norrgard, another long-term Western Fuels associate. WFA is a cooperative of coal-dependent utilities in the western states that works in part to discredit climate change science and to prevent regulations that might damage coal-related industries. Their main message is that CO_2 emissions are good for the planet; coal is the best energy source we have. Cooperating individuals are Patrick Michaels, Robert Balling, David Wojick, Sallie Baliunas, Sylvan Wittwer, John Daley, and Sherwood Idso. They receive their funding from the WFA, which in turn receives its funding from its coal and utility company members.

ExxonMobil is the largest publicly traded corporation with 2005 revenues of $339 billion. Just this single corporation has funded at least 43 organizations to the tune of about $16 million between 1998 and 2005 to spread disinformation on climate change.

The Center for the Study of Carbon Dioxide & Global Change claims to "disseminate factual reports and sound commentary on new developments in the world-wide scientific quest to determine the climactic and biological consequences of the ongoing rise in the air's CO2 content." The Center is led by two brothers, Craig and Keith Idso. Their father, Sherwood Idso, is affiliated with the GES, and the Center also shares a board member (Sylvan Wittwer) with GES. Both

Idso brothers have been on the Western Fuels payroll at one time or another. Their main message is that increased levels of CO_2 will help plants, and that is good. The Center's funding sources are extremely secretive, stating that it is their policy not to divulge it funders. There is evidence for a strong connection to the GES (ergo WFA). Connected individuals are Craig Idso, Keith Idso, and Sylvan Wittwer [12].

On a different scale and intellectual level are Dr. Richard Lindzen, a legitimate scientist, and Alfred P. Sloan, professor of Atmospheric Science at the Massachusetts Institute of Technology. He is considered a contrarian on climate change and is constantly picking at particular points without standing back and looking at the overall situation. Dr. Lindzen is a meteorologist who for years has expressed skepticism about some of the more dire predictions of other climate scientists about the significance of human-caused warming. He receives support for his staff to publish these contrarian views directly from the coal industry. One of his typical articles can be found in popular media [13].

However, Dr. Lindzen was a member of the prestigious National Academy of Science panel that reviewed climate change early in the Bush administration [14]. This report was written by 11 atmospheric scientists who are members of the Academy. They concluded that, "the IPCC's conclusion that most of the observed warming of the last 50 years is likely to have been due to the increase in greenhouse gas (GHG) concentrations accurately reflects the current thinking of the scientific community on this issue. Despite the uncertainties, there is general agreement that the observed warming is real and particularly strong within the past 20 years." Thus, when he does stand back and view the overall situation, even Prof. Lindzen does support the consensus view of the rest of the science community. However, when not involved with other scientists and when not submitting peer-reviewed articles, he has been heard to make remarks in public like, "cigarette smoking does not cause cancer and global warming is due to natural phenomena" (personal conversation with Professor Jeff Severinghaus at the Scripps Institute of Oceanography at UCSD on June 11, 2008). Why would an excellent scientist say such things that he knows are not true? Good question. You would have to go beyond science and ask a few questions about his political ethics. Let us assume that he believes that government regulation is the greatest wrong. And let us assume that telling a truth (cigarette smoking does cause cancer and global warming is caused by human activities) will increase the chances of government regulation. Then, it is quite reasonable that telling an untruth publicly is a lesser wrong if it reduces the chance of having government regulation in this area. Well, these turn out to be pretty good assumptions (personal conversation with Professor Jeff Severinghaus at the Scripps Institute of Oceanography at UCSD on June 11, 2008).

These energetic and well-financed groups have achieved a remarkable record of having about half of the articles even in major newspapers (*New York Times*, *Washington Post*, *LA Times*, and *Wall Street Journal*) support the claim of doubt about the cause of climate change. You really have to take off your hat to this dedicated cabal. They certainly rival the long-term success of the gifted writers who were able to cast doubt on the relationship between cancer and smoking cigarettes. Their publishing in the popular media fits well with a reporter's typical approach to a story of getting statements from both sides of an issue. Reporters are not scientists and are apparently vulnerable to these ill-formed opinions by those on the "other side" of the climate change issue.

There is essentially no debate in the political arena about human-caused climate change with 165 countries ratifying the Kyoto Protocol to date with only one developed nation abstaining—the United States through the second Bush administration. Climate change is considered by almost all of the world's nations to be a serious problem. At this point in the evolution of the Protocol, 35 industrial countries have CO_2 levels specified for each of them in the treaty.

Many of these nations have expressed demoralized sentiments about their efforts when the major polluter (the United States) has, for 14 years, shunned this international attempt to deal with global heating. The 2-week UN conference on climate change in Nairobi with 6000 participants ended in late 2006 without agreeing on how to move beyond the Kyoto Protocol. The current protocol only limits the emissions by most industrialized countries, but this expires in 2012. The two persistent problems cited at the conference were U.S. reluctance to agree to any mandatory emissions limits and the increased stubbornness by China and India to participate in the next phase of the Protocol without U.S. involvement. Delegates expressed growing frustration with the Bush administration saying that without clear signals from the world's largest source of air pollution, other countries would hesitate to move ahead [15].

3.2   WHAT IS WRONG WITH 5 DEGREES FAHRENHEIT?

Climate change was first suggested by Svante Arrhenius in Sweden in 1896. He proposed that existing atmospheric gases such as CO_2, methane, and nitrous oxide would be augmented by industrial-era gas generation. These higher levels of GHG could cause warming of the planet. To these naturally occurring gases are added the other GHGs such as sulfur hexafluoride, perfluorocarbons, and hydrofluorocarbons, which come only from human sources. It should be noted that at higher temperatures, there is increased water vapor in the atmosphere and this is also a heat-trapping gas. Since Arrhenius lived in Sweden with severe winters, he thought that this warming would not be such as bad idea.

Thus, the very first person discovering global warming as a scientific phenomenon had misunderstood the potential implications of his own work. The rest of us were also slow to understand the implications and the severity of the continued pumping of a CO_2 and these other GHGs into the atmosphere. This view of the mildness of this human-driven climate change was reinforced by the very phrase *global warming*—such a comforting term. The numerical quantization of the warming effect was measured in degrees Celsius and numbers like 0.74° were bandied about. What could be so bad about 0.74° here or there? It is with ever-widening eyes that we view the future that is likely to emerge based on our current global experiment of pumping 7 GtC/year (billion tons of carbon) into the atmosphere every year. (This is equal to about 26 $GtCO_2$/year.)

The earth's dynamics and the likely results of increased GHGs are very complicated subjects. It has taken decades of dedicated work by thousands of scientists to garner a clearer understanding. Currently, this complexity is managed by about 20 detailed computer simulations being conducted for a particular set of assumptions of future human actions. These resulting data sets are then used as inputs to several global climatic computer models, each of which specialize in certain aspects of the modeling such as the ocean, the land, or the atmosphere. A typical simulation takes hundreds of hours running using a supercomputer for each climate model. These results are then combined into a more holistic understanding of the global situation. As it turns out, over the last 10 years, almost every new scientifically established insight found outside the laboratory and used in this complex network

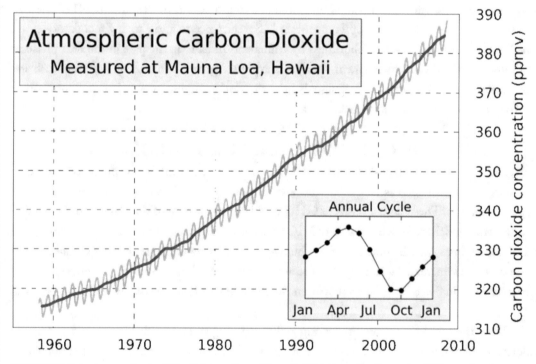

FIGURE 3.1: Atmospheric CO_2: Mauna Loa, Hawaii. This is a file from the Wikimedia Commons. Image: Mauna Loa Carbon Dioxide.png, uploaded in Commons by Nils Simon under license GFDL & CC-NC-SA; itself created by Robert A. Rohde from NOAA published data and is incorporated into the Global Warming Art project.

of computer simulations has worsened our projections of what the likely outcome is of human-driven global climate change!

The experimental science really started with the first direct global CO_2 measurements initiated in 1958 by Roger Revelle and carried out by Charles Keeling at a test station in Hawaii. The resulting plot of the concentration of CO_2 in the atmosphere over time is called the Keeling Curve. The CO_2 levels were already at 310 parts per million (ppm), which is above the preindustrial level of 280 ppm established later by ice core sampling. Currently, the CO_2 level is at 385 ppm, which is about 100 ppm above our preindustrial starting point and climbing as shown in Figure 3.1.

There is ample evidence that the expected earth warming is a physical reality. The average earth temperature rise over the last hundred years was 1.33°F or 0.74°C with most of the increase occurring since 1976 (see Figure 3.2). The projection from 1990 to 2100 is for a 3- to 8-°F average global temperature rise (IPCC Climate Change 2007: The Scientific Basis Working Group 1 Contribution to the Third Assessment Report). This average global increase is averaged over the entire earth's surface, both water and land. The increases over water are less than average and those over land will be more than average. This average global increase implies up to 15° of increase locally on land. Things are accelerating. Eleven of the past 12 years rank among the dozen highest temperatures recorded since the mid-1880s.

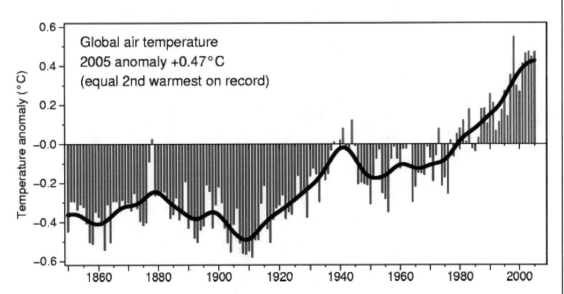

FIGURE 3.2: Global air temperature: 2005 anomaly +0.47°C (equal second warmest on record). University of East Anglia, U.K. Multiply by 1.8 to convert from degree centigrade to Fahrenheit.

The CO_2 level in the atmosphere and the global temperature correlate directly. This period around in the historical cycle is normally called the "ice ages"—as the CO_2 goes up, the temperature goes up. Measurements taken over the last 650,000 years using ice core samples in the Antarctica show that CO_2 levels have varied over a maximum range between about 180 and about 275 ppm. The corresponding temperature change over this historical range of 100 ppm caused earth conditions to go from a mile-thick ice in a mid-latitude location such as Chicago to the current post-ice age temperatures with Chicago being hot and muggy in the summer. So about 100 ppm of CO_2 is a big deal on this planet. At no time in the last 650,000 years has the CO_2 level exceeded 300 ppm (see Figure 3.3).

The natural variability in the CO_2/temperature oscillations in the past during the cyclic ice ages was driven primarily by the small variations in the earth's orbit (the wobble of the earth's axis, the tilt of the earth's axis, and the idiosyncrasies of the earth's orbit). When you consider the three small effects, you can deduce the resulting deviation in the shape of the earth's orbit and tilt toward the sun. The sum of these small effects is the driving force behind past variation in earth's CO_2 and temperatures that has produced ice ages and warm interglacial periods. The earth heated up as it drew nearer the sun for long periods and the earth's temperature rose. For about 800 years

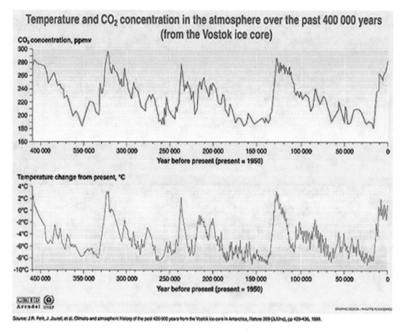

FIGURE 3.3: CO_2 levels today are higher than over the past 420,000 years. From J.R. Pettit, "Climate Atmospheric History of the Past 4200 years from the Vostok Ice Core in Antartica", Nature, pp. 429–436, June 3, 1999.

during this warming, the CO_2 levels typically remained steady. At this point, the increasing ocean temperature caused the ocean to release CO_2 because warmer water holds less CO_2. For the rest of the interglacial warm period, both the orbital mechanics and the rising CO_2 caused the earth's temperature to rise for thousands of years. Overall, the CO_2 contributed about 30% of the overall temperature rise and the earth's orbital mechanics did the rest.

The current CO_2/temperature episode is not driven by slight changes in orbital mechanics but by human activities rapidly dumping excessive amounts of GHGs into the atmosphere. The GHGs are driving the temperature increases above and beyond natural phenomena.

How do we know it is not naturally caused CO_2 since human-generated CO_2 is a small fraction of the total natural CO_2 release? It turns out that natural CO_2 is in fact a different gas than the human-released CO_2 primarily from burning fossil fuels. Natural CO_2 is a mixture of three types of CO_2: CO_2 (isotope 12), CO_2 (isotope 13), and CO_2 (isotope 14), whereas human-generated CO_2 is composed of only CO_2 (isotope 12). By simply measuring the relative amounts of the different isotopes of CO_2 in the atmosphere, you can tell where the CO_2 came from—natural or human sources. This is one of the key indicators used by the IPCC scientists to determine what is driving the current episode of global temperature increase. When these computer simulations are run, good correlation with historic temperature data is achieved only when anthropogenic (people) forcing is added to natural variability as shown in Figure 3.4.

Now at 385 ppm CO_2, we are about 100 ppm above the preindustrial level of 280 ppm. This large CO_2 rise is extraordinary when the conditions over the last 650,000 years are considered. CO_2 levels are expected to reach about double the preindustrial level of 560 ppm by 2050 and will continue to rise throughout the 21st century if we continue in our business-as-usual approach to our energy use. This is about twice the difference in the CO_2 level that caused the 1 mile of ice over Chicago to melt in today's temperatures.

With the business-as-usual use of fossil energy continued with a 2% per year increase in CO_2 emissions, then the annual growth of CO_2 concentration in the atmosphere would increase to 4 ppm/year by 2050. This extra CO_2 would lead to a global warming of at least 5°C by 2100. Let us put this possibility in perspective. Global warming of 1°C more than the year 2000 temperatures would be a large climate change and 3°C would yield a "different planet" [16]. A 5°C increase would be ???

The IPCC baseline projection of about 45 cm (18 inches with a range of 10–90 cm) ocean level rise by 2100 assumes no contribution from Greenland or Antarctica ice melts. This represents only the thermal expansion of the oceans as it gets hotter. If these high levels of CO_2 persisted so that the planet's innate thermal inertia would catch up with this higher temperature level, then both Greenland and Antarctica would melt. The last time the world's CO_2 level actually stabilized at 500 ppm was 20–40 million years ago. The ocean levels at that time were 300 feet higher than

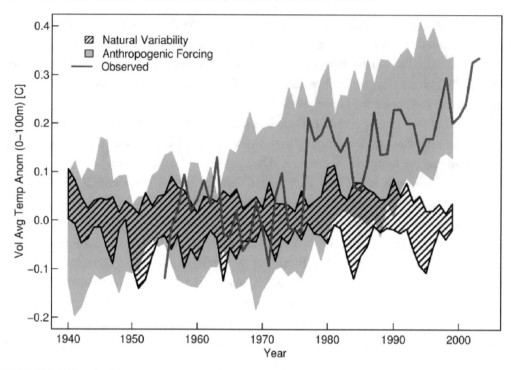

FIGURE 3.4: People driven computer simulations correlate well with observed upper-level ocean temperatures, 1940–2003. (David W. Pierce, Scripps Institution of Oceanography.)

today, and it was a different planet. There were palm trees above the Arctic Circle [17]. Thus, CO_2 concentration levels above 500 ppm are really big deal!

This climate variability has been studied using increasingly sophisticated computer models. Current trends are well beyond natural variability and cannot be explained by computer models just based on natural causes. Only when natural and human causes are considered simultaneously can the recent temperature record be understood.

The impacts of this climatic disruption of our planet's ecosystem are already proving legendary. Excessively high temperatures in 2003 cause more than 35,000 deaths in Europe alone. This large loss of life was in developed countries! There were severe losses in many other countries that are not as well recorded. The intensity of tornadoes, hurricanes, typhoons, and cyclones is increasing [18] with an increasingly high toll in loss of life and property damage. New to this list of phenomena is the unheard-of occurrence of a hurricane in the South Atlantic. The increased intensity of storms is directly linked to the heating of the oceans. The monetary damages caused by weather and flood catastrophes have increased by a factor of 20 in the last 45 years. The number of major flooding events worldwide has increased from 78 to 630 over the last 50 years—by a factor of nearly 8 [11].

In parallel with increase flooding is a significant increase in droughts. The amount of desertification has more than doubled over the last 20 years to about 2000 square miles per year (size of the state of Delaware). Although overall precipitation has increased by about 20% over the last century, the higher temperatures cause accelerated drying out of land. This has caused persistent drought in areas such as most of Africa especially sub-Saharan, southern Europe and Turkey, parts of central Europe, parts of central and southern Asia, parts of China, southwestern South America, parts of the Caribbean, and parts of the Western and Southwestern United States. Business-as-usual use of energy would mean that in less than 50 years, the loss of soil moisture could be as much as 35% in large growing areas of the United States. Combining this with the current practice of overdrawing underground water reservoirs and you have a perilous situation [19].

So far, we have discussed the average global temperature rise. The magnitude of the temperature increase is much larger at the poles. For example, an average global temperature rise of 5° may only increase the equatorial temperatures by 1°. However, this will raise the polar temperatures by 12°.

The implications of this variance in thermal effects are profound. This polar magnified effect will cause the polar ice caps to melt at an accelerated rate. These polar phenomena seem to be happening at rates that are beyond recent expectations of scientists. The Artic polar ice size has been reduced from about 13,500,000 km² around 1900 to the current level of 11,800,000 km², which is a decrease of 13%. This loss of Arctic ice is more than half the size of India (see Figure 3.5).

In addition, the Artic summer sea-ice thickness has decreased by an estimated 40% since the 1950s. It is projected to disappear completely during the summer within the next 50 years. This ice reduction will accelerate the absorption of sunlight in the polar regions because ice reflects back 90% of the incoming sunlight, whereas open water reflects only 10%. This is an example of a negative feedback mechanism that is not self-correcting and makes the heating phenomena accelerate.

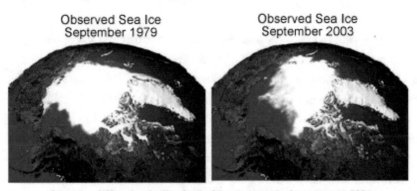

Impacts of Warming Arctic, *Arctic Climate Impacts Assessment*, 2004

FIGURE 3.5: Arctic ice cover reduction in 24 years.

Also important is the recent finding that CO_2 and methane trapped in frozen tundra are being released as the tundra melts. The number of days each year that tundra is frozen enough for vehicle travel in Alaska has reduced from 220 days per year in 1970 to 100 days per year today. As the tundra-trapped CO_2 and methane are released, they increase the GHG loading in the atmosphere especially since methane has 25 times the green house warming effect of CO_2. This is another negative feedback mechanism that accelerates global heating. This is happening along the southern edge of the tundra zone around the entire northern hemisphere. This extra CO_2 emission has not yet been included in climate models.

Temperate zone frost days are decreasing at a steady rate. One location in southern Switzerland now has only 15 frost days, whereas it was around 70 days as recent as 1950. This is happening in a zone that circles the globe in both the northern and southern hemispheres. Among other things, this allows invasive species to move north, and their number has increased dramatically from 2 to 14 over these 50 years in this single location in Switzerland. This warming allows disease vectors to move into new areas. Malaria-carrying mosquitoes are set to expand their range dramatically into northern Europe and the highlands of tropical Africa.

Another such invasive species is the bark beetles that kill drought-stressed trees and provide the dry fodder for increasingly severe forest fires. These bark beetles have killed more than 14 million acres of spruce trees in Alaska and British Columbia alone. The number of major wildfires in the Americas has increased from 2 per decade to 45 per decade in the last 50 years. Southern California, which is currently infected with these same bark beetles, had the second most severe fire in California-recorded history in 2003 when the Cedar Fire burned 721,791 acres (2921 km²), which is about the total land area of Rhode Island. This single fire destroyed 3640 homes (including the home of the author) and killed 15 persons. The estimated value of the destroyed property was about $2 billion in this single fire (see Figure 3.6). It was one of 15 fires raging at the time in southern California. At this moment, there at 1200 fires raging in northern California.

The term *forest fire* is being replaced by the phrase *fire storm* because of the severity and intensity of these recent fires.

Coral reefs are vital to ocean species. Warming has contributed to coral damage by a combination of higher temperatures and excess CO_2 absorbed in the oceans. The oceans are the major sink for the extra CO_2 in the atmosphere. Between 1800 and 1994, fossil fuel use and industrial processes such as cement production generated some 900 billion metric tons of CO_2. During this time, the atmospheric concentration of GHGs rose from about 280 to around 360 ppm. The world's oceans absorbed about half (433 billion metric tons) of this industrial CO_2 that was emitted [20].

The threat to coral follows from the carbonic acid that forms from the dissolved CO_2 and increases the acidity of the water. This acidity increase changes the saturation level of calcium carbonate. This damages not only coral but many other small sea animals that use this chemical as their basic building block [21].

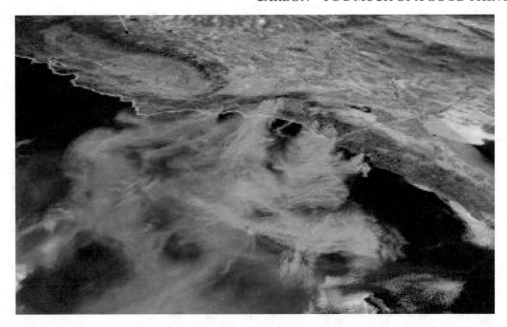

FIGURE 3.6: Smoke from cedar fire in San Diego in October 2003.

The fraction of temperate and equatorial oceans that has extremely low levels of calcium carbonate has changed from 0% before 1880 to 25% today. Estimates based on business-as-usual projections of energy use indicate that approximately 60% of this important section of the oceans will have unacceptable levels of calcium carbonate in the year 2050. The food chain in the bottom of the ocean is being threatened, and this will have dire impact on ocean protein that millions of people depend on [11].

There are enormous ice packs sitting on land in Greenland and west and east Antarctica. It startled scientists in early 2002, when the 150-mile-long and 30-mile-wide Larsen-B ice shelf broke up in about 1 month. They had thought that this mass of ice floating on the ocean but attached to Antarctica would be stable for another century (see Figure 3.7 of Larsen-B after the breakup).

Equally startling was the abrupt breaking away of a 25-square-mile shelf of floating ice from the north coast of Ellesmere Island in northern Canada. This 100-foot-thick slab was part of 3900 square miles of ice shelves that existed in 1906 when the Arctic explorer Robert Peary first surveyed the region. Ninety percent of this ice is now gone [22]! "The quick pace of these changes right now is what stands out," said Dr. Luke Copland, the director of the University of Ottawa's Laboratory for Cryospheric Research.

A recent scientific finding has uncovered a mechanism that is accelerating the movement of the ice into the oceans. Meltwater on the top of the ice fields was thought to refreeze as it passed down into the base ice. It was a surprise to find that this water courses down through the ice

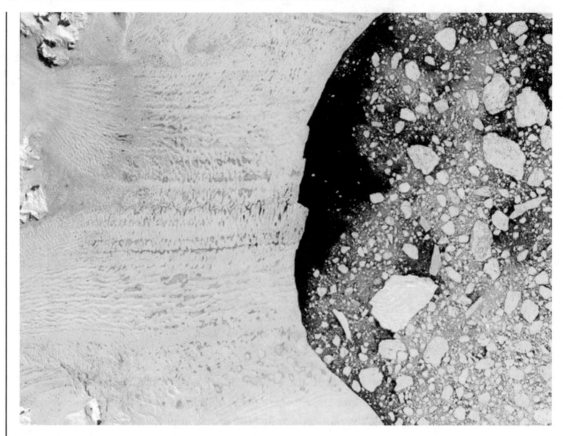

FIGURE 3.7: Rapid break up of the Larsen-B ice shelf in Antartica in 2002.

creating many vertical channels or a Moulin and the water can reach the base of the ice at the bed-rock. This liquid lubricates the mass of ice above it and expedites the movement of the ice into the ocean.

As eluded, this accelerating melting of polar ice will directly contribute to another impact of climate change—the rising of ocean levels. At this point, the ocean level has risen between 4 and 10 inches caused mainly by the expansion of the ocean due to temperature rise. This is 10 times higher than the average rate during the last 3000 years. The amount of ice flowing into the sea from large glaciers in southern Greenland has almost doubled in the last 10 years. This is causing scientists to increase estimates of how much the world's oceans could rise under the influence of global warming [23].

An early indicator of this effect is the closures of the Thames River barrier built to protect London from extreme tides. Built in the 1930s, this sea level barrier was opened two or less times a year for 50 years after it was built. Currently, it is being opened up to 11 times per year and is an early indicator of rising ocean level [11].

If the ice covering Greenland melts, the oceans will rise 23 feet. A similar amount of ice is on the west Antarctic ice shelf, and its melting will cause another 20-foot rise in sea level. The much larger east Antarctic ice shelf could add even more water to the ocean level. Temperature rise at the pole is magnified by a factor of 10 more than at the equator. This focusing of global heating at the poles will accelerate the ice melting. The Greenland and Antarctic ice will melt eventually with the business-as-usual future. The only question is how much of this will happen by 2100 for a frame of reference for the near term. However, although there is no doubt about the final outcome, the uncertainty for the IPCC is over how fast Greenland's ice sheet will melt. The rates of all the melting processes are still very uncertain, nor are they properly represented by conventional models for predicting melting rates.

Without deviation from the business-as-usual energy use, polar ice melting will proceed unabated and eventually reach a rise in the ocean level of 10, 20, and 40 feet or more. If the CO_2 stabilized at 500 ppm and was not reduced back to near-preindustrial levels, the expected rise in ocean levels would be 300 feet based on data from about 30 million years ago. Current rough estimates are that this would take 200–1000 years if CO_2 levels are not brought back to preindustrial levels. The civilization we have built over the last millennia would have to abandon coastal cities and move somewhere inland. This is Katrina on a large scale.

A 20-foot ocean rise would submerge large coastal land areas where people tend to form settlements. This will cause enormous economic losses and the dislocation of hundreds of millions of people. Some of the areas especially impacted would be Florida, New York City, San Francisco, New Orleans, London, the Netherlands, Beijing, Shanghai, Calcutta, Bangladesh, the Pacific Island nations, and many other coastal areas around the world. The Sacramento River delta dike system, which is already 13 feet below sea level, would be overwhelmed, disrupting the water supply to southern California.

The number of refugees from global heating would swell even further when the multiple effects of drought, flood, ocean rise, intense ocean-borne storms, and disease vector movements into new areas all afflicting the planet in tandem. The magnitude of this collective impact would be staggering.

The U.S. Defense Advanced Research Projects Agency (DARPA), which does advanced development and thinking for the U.S. military, sponsored a recent study on one possible effect of rapid climate change induced by global heating [24]. The Pentagon's Andrew Marshall asked researchers to examine the likely repercussions of a disturbance of the circulating ocean currents. DARPA was primarily interested in the impacts on the U.S. military ability to deal with such a climatic event. Peter Schwartz, former head of planning for Shell Oil, and Doug Randall of the Global Business Network, a California think tank, conducted the study.

The rapid climate change mechanism that the researchers choose to study was the interruption of the "global ocean conveyer belt." This circulating ocean current winds its way through all the oceans on the planet. It is the planetary force that provides stability to our current global climate, and among other things, warms northern Europe. One section of the belt, the Gulf Steam, brings warm salty surface water from equatorial regions to the north across the Atlantic Ocean to the western coast of Europe. This water is cooled in these northern regions, and gets heavier and sinks to the bottom of the ocean at the staggering rate of 5 billion gallons a second, which is the equivalent of 65 Mississippi rivers. It is this dropping of the heavier water in the North Atlantic and Pacific that powers the global circulation system. The current then moves southward under the Gulf Stream and is part of a globe-spanning circulation current—the thermohaline circulation.

This ocean current was interrupted toward the end of the last ice age when meltwater trapped in lakes in North America behind the moraine formed by the ice age glaciers broke through and flowed down what became the St. Lawrence River into the northern Atlantic. The Chesapeake Bay was formed by a similar occurrence. This cold fresh water was lighter than the warm salty Gulf Stream water and floated on top, diluted, and interrupted the Gulf Stream flow. This event disrupted global climate and caused the reimmersion of the northern Europe into the ice age. This transition to ice age happened in a very short period like a decade and the effects lasted a millennium.

In our current era, this could happen by fresh water runoff from melting ice from Greenland and high-latitude precipitation. This fresh water would dilute the salty warm water in the North Atlantic and this lowered density would in turn pave the way for a sharp slowing of the thermohaline circulation system. A second possibility is the hot surface water, which is part of this global circulation that circles Antarctica, being diluted by increased rain and ice field melt.

The DARPA-sponsored study concluded that if the ocean circulation system stopped, the likely results of this climatic event would overwhelm the U.S. military (among other institutions). This was due to the projected rapid cooling of the northern hemisphere up to 5.5°F in one decade. There would be disruption of rainfall pattern and the resulting intensification of floods and droughts altering agricultural pattern relatively rapidly. The earth's carrying capacity would be exceeded causing desperate strife and causing mass migration of hundreds of million if not billions of people. This would tear the fabric of the civilized world and result in total chaos. The military would be overwhelmed in attempts to provide stability.

Because past changes in thermohaline circulation have occurred during periods of relatively rapid climate change, the rapid warming we are currently experiencing could trigger an abrupt thermohaline shutdown and subsequent regional cooling. Although a shutdown of thermohaline circulation is unlikely to occur in the next century, scientists have recently found that freshwater inputs have already caused measurable "freshening" of North Atlantic surface waters over the past 40 years. Human activities may be driving the climate system toward a threshold and thus increasing the chance of abrupt climate changes occurring. Some scientists have reexamined this possibility in light of the accelerating melting of polar ice especially from Greenland.

A recent study published June 17, 2005, in *Science* by Ruth Curry of the Woods Hole Oceanographic Institution and Cecilie Mauritzen of the Norwegian Meteorological Institute, examined the magnitude of fresh water addition and the impacts on the North Atlantic section of the thermocline. They conclude that uncertainties remain in assessing the possibility of circulation disruptions, including future rates of greenhouse warming and glacial melting. Most computer simulations of greenhouse warming show increased precipitation and river runoff at high latitudes that leads to a slowing of the Atlantic conveyor. Only one available computer model study, however, contains an interactive Greenland Ice Sheet. Pooling and release of glacial meltwater, collapse of an ice shelf followed by a surge in glacier movement, or lubrication of the glacier's base by increased melting are all mechanisms that could inject large amounts of freshwater into the critical upper layers of the Nordic Seas.

These cold, dense southward flowing waters (collectively called Nordic Seas Overflow Waters) have been closely monitored with instrument arrays for more than a decade, but no sustained changes have yet been measured. Of the total 19,000 km^3 of extra fresh water that has diluted the northern Atlantic since the 1960s, only a small portion—about 4000 km^3—remained in the Nordic Seas; and of that, only 2500 km^3 accumulated in the critical layer feeding the Overflow Waters. (For reference, the annual outflow from the Mississippi River is about 500 km^3.)

This is the reason, Curry explains, why the overflows have not yet slowed despite all the freshening. At the rate observed over the past 30 years, it would take about a century to accumulate enough freshwater (roughly 9000 km^3 according to this study) in the critical Nordic Seas layer to significantly slow the ocean exchanges across the Greenland–Scotland Ridge and nearly two centuries of continued dilution to stop them. The researchers conclude therefore that abrupt changes in ocean circulation do not appear imminent.

"It certainly makes sense to continue monitoring ocean, ice, and atmospheric changes closely," Curry said. "Given the projected 21st century rise in GHG concentrations and increased fresh water input to the high latitude ocean, we cannot rule out a significant slowing of the Atlantic conveyor in the next 100 years."

CO_2 is the major GHG, but it is not the amount of the gas that is important but its contribution to greenhouse warming. For example, methane (CH_4) is 25 times more potent than CO_2 in its planetary warming effect. The convention is to measure this warming in tons of CO_2 equivalent to put all GHGs on the same footing. When measured on this scale of equivalent warming effect, human-generated CO_2 is 84% of the total warming. Methane is next at 9% and the other five GHG of note produce the remaining 7% [25]. Focusing on the major GHG, CO_2 in the United States comes from oil combustion (42%) and coal use (37%). The oil is used primarily in vehicle transportation (about 65%), and coal is used almost exclusively in electricity generation (90%). Although coal generates only 51% of U.S. electricity, it is responsible for 81% of the GHGs in the electric sector. These two uses, oil in vehicles and coal in electricity, account for 79% of the U.S. GHGs dumped into the atmosphere. Any strategy to reduce global heating must directly address these two primary sources of CO_2. When we look at end uses of energy, GHGs are distributed so that transportation accounts for 27%, industrial uses are at 39%, and residential/commercial buildings release 34%.

CO_2 has the unfortunate characteristic of remaining in the atmosphere for a long period and the excess carbon is first diluted by the carbon cycle as it mixes into the oceans and biosphere (e.g., plants) over a period of a few hundred years. Then, it is slowly removed over hundreds of thousands of years as it is gradually incorporated into carbonate rocks. About 25% of the CO_2 will remain in the atmosphere after about 200 years. If we completely stopped all CO_2 generated this minute, the existing amount of CO_2 in the atmosphere will continue to drive the planet's climate change for most of the next century.

Estimates of the maximum level of atmospheric CO_2 to keep onerous impact from occurring are difficult to make, but there seems to be some agreement that 450–550 ppm is the limit. Beyond this level, most researchers who have examined this question have indicated that climatic instabilities would be increasingly unacceptable. This is not a black-and-white estimate, but it seems to be a reasonable value to base policy on given our current understanding of this complex issue. Five hundred fifty parts per million represents a doubling of the preindustrial level of CO_2. It is nearly three times the 100-ppm difference that caused mile-thick glaziers over most of the temperate zone during the ice age to melt to today's post-ice-age temperature.

A growing number of governments and experts have concluded that the world's long-run temperature rise should be limited to 2 or 3°C above the current level. The CO_2 level should be stabilized at 450–550 ppm to achieve this limit to temperature increase. This is twice the preindustrial level—an increase of up to 270 ppm. Recent estimates by Jim Hansen indicate that 450 ppm is a more realistic goal to avoid the worst of the impacts of climate change. In December 2007 at the American Geophysical Union meeting in San Francisco, Dr. Hansen announced the results of his latest study of the paleoclimatic data. This led him to conclude that unless CO_2 is stabilized under 350 ppm, Earth's permanent ice sheets would melt completely. This result is not yet accepted by the IPCC community of scientists. If born out by the elaborate IPCC process, the 2050 goal of 450–550 ppm for this study may be inadequate.

To achieve a global peak of 450–550 ppm CO_2 in the atmosphere means that we have to collectively reduce our current CO_2 generation to about 50% of current value by the year 2050. Hansen estimates that this goal for the globe should be a 75% reduction in carbon emission level [26].

This is indeed a stiff challenge to our technical ingenuity, political wisdom, and international cooperation. For all this to take place, we face and additional challenge to change our personal way of doing things. To change the business-as-usual energy trajectory into the future, we must change our personal business-as-usual now.

The only good news in this almost overwhelming planetary crisis is that global oil production is peaking this decade. The oil peak challenge is to avoid overdemand driving up oil prices and causing global economic chaos by collectively reducing our global demand in line with diminishing production. Then, this collective action will actually contribute to the stringent goal of reducing CO_2 by 2050. If we deal with the oil peak, it will help us deal with the CO_2 overburden.

The United States has generated enormous bad will because of our opposing the implementation of the Kyoto Agreement and steadfastly refusing to even discuss the climate change problem at international meetings. The United States is "the" major historic global warming gases polluter. Over the couple of hundred years of the industrial era, the United States was the major polluter generating 30.3% of all GHG followed by Europe (27.7%), Russia (13.7%), and Japan (3.7%). Together, this handful of countries contributes three quarters of today's GHGs now present in the atmosphere. The industrial and developed countries of the world cause the historic and current CO_2 problem. It is up this small number of countries to lead the way in solving the problem. Eventually, the developing world, especially China and India, must be drawn into this international effort to restore stability to the planet's climate. At this point, they refuse to participate until the United States joins the effort.

Scientific evidence suggests we must prevent global average temperatures from rising more than 2°C (3.6°F) above preindustrial levels to avoid the worst effects of climate change. To stay below this threshold, atmospheric concentrations of CO_2 must stabilize at or below 450 ppm. This goal requires cutting today's levels of worldwide climate change emissions by 50% to 75% by mid-century with additional cuts later in the century. Given that the United States leads the world in both absolute and per capita emissions, we must achieve even deeper reductions here at home, on the order of 80% below 2000 levels by 2050. This leads us to consideration of what the carbonless options are and how much of a role can they play in achieving this very difficult goal.

· · · · ·

CHAPTER 4

Carbonless Energy Options

Coping with the near-term prospect of the peaking of global oil production and the need to significantly reduce greenhouse gases (GHGs) presents a formidable challenge to the peoples on the planet earth. Because the United States is the major user of oil and the largest historic producer of GHGs, it is instructive to develop a reasonable strategy for America. The other industrialized countries have already committed to the Kyoto Protocol, and the task seemed daunting without the involvement of the major polluter. When the United States takes a responsible approach to reducing its carbon emission, this action will further stimulate and strengthen the commitment of all those already dedicated to this purpose. With all the industrialized nations involved, it would finally lay the basis for future talks with the developing nations such as China and India, and to eventually include them in this monumental task to save the planet from toasting.

Exceeding global average temperature increases above 2–2.5°C above the 1750 preindustrial level would entail "sharply increasing risk of intolerable impacts" [27]. To avoid triggering unprecedented warming with potentially disastrous consequences, the people of the planet collectively have to limit the additional temperature rise due to greenhouse gases to 1°C above today's level (Jim Hansen speaking to ASES National Conference, Denver, 2006). To avoid exceeding this temperature limit will require stabilizing atmospheric concentrations at the equivalent of no more than 450–550 ppm of carbon dioxide (CO_2) (compared to about 380 ppm CO_2 equivalent today). That in turn requires that global CO_2 emissions peak no later than 2015 to 2020 at not much above their current level and decline by 2050 to about a fifth of that value.

The global total emissions of carbon are currently 7 GtC/year, which is 7 billion metric tons of carbon per year. (To convert from tons of carbon to tons of CO_2, you would multiply by 3.7 times.) Today, the handful of industrialized nations is responsible for about half of the world's total carbon emissions. Historically, the industrialized countries have contributed the lion share of CO_2 since the start of the industrialized era.

There is a strong equity issue surrounding the carbon emission issue. The economically successful countries have created the magnitude of the current global warming problem over the last 2.5 centuries. The underdeveloped countries, because of their poverty, have not been part of generating this global problem. The needs of the countries just on the cusp of economic development such as

China and India, and the other underdeveloped countries that hope someday to rise out of poverty, need to be considered. Without consideration of these needs, there will be little basis for the future negotiations about including them in the Kyoto Protocol. To create the space for their growth and rise from poverty, the developed countries need to make a greater contribution to carbon reduction.

To keep the atmospheric CO_2 under the 450–550 ppm equivalent limit, the carbon dumping needs to be reduced in the United States by about 80% by 2050. Today's U.S. carbon deposition in the atmosphere is 1.6 GtC/year. To be on track, we need to reduce the carbon released by about 40% by 2030 and the U.S. total carbon emission goal at that time would be about 1.0 GtC/year.

The U.S. business-as-usual (BAU) growth in carbon use, which is about 1.4% annually, would bring us to 2.2 GtC/year by 2030 to account for population and expected economic growth [US Department of Energy (DOE) Energy Information Agency (EIA)]. The United States would need to reduce its carbon use from this projected 2.2 GtC/year to the 1.0 GtC/year goal. As shown in Figure 4.1, this is a reduction of 1.2 GtC/year between now and 2030 to be on track to the 2050 goal.

This is about as stringent as the Kyoto protocol agreement which states that most industrialized countries will reduce their collective emissions of GHG by 5.2% compared to the year 1990. Compared to the emissions levels that would be expected by 2010 without the Protocol, the Kyoto agreement represents a 29% cut.

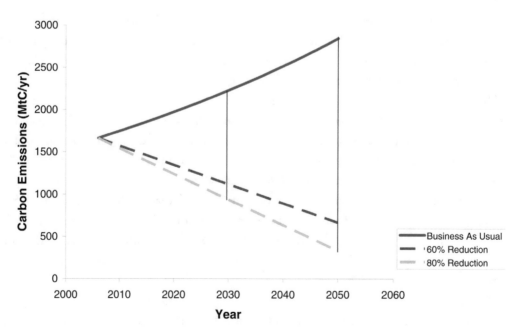

FIGURE 4.1: U.S. fossil fuel reductions needed by 2030 for a 60% to 80% reduction from today's levels by 2050.

To investigate the possibilities of more directly meeting this goal of reducing U.S. carbon emissions by 1.2 GtC/year by 2030, the potential of increasing energy efficiency and introducing renewables were examined first. To do this, 18 researchers were approached who are expert in nine technical areas and they were asked to establish what could be done in each of their areas. The strategy developed relies on both increasing energy efficiency in a cost-effective manner in three primary areas and introducing greater amounts of renewable energy in six technology areas. A bottom-up approach is taken with each expert asked to come up with their best estimates of what can reasonably be done in their area considering resource limits, economics, and existing barriers to introduction [28].

Because of the seriousness of the problems, it is not of much use to do a blue sky analysis that touts extreme possibilities in each area. What is needed now is a hard, cold look at each avenue and to make projections based on competitive economics as well as reasonable restraints to include difficulties and barriers that may exist for each approach. Each of the nine areas are examined separately and then put together to see how close this assembly of options comes to reaching the needed goals. This effort represents both an attempt to avoid hitting the wall of the oil production peak and that of providing the energy services to a growing society while significantly reducing the carbon dumped into the atmosphere.

Buildings, transportation, and the industrial sectors were examined for cost-effective energy saving opportunities; and detailed studies were completed looking at how electricity, natural gas, and petroleum reductions would be achieved. In addition, wind, photovoltaics (PVs), concentrating solar, biomass, biofuels, and geothermal are the renewable energy technologies considered. Depending on the results of these nine carbonless opportunities, there are two additional technologies that could be considered and they are "clean coal" and "safe nuclear." An example of "clean coal" would be an integrated gasification combined cycle (IGCC) coal plant with carbon capture and sequestration. This would increase the efficiency of generating electricity from coal and significantly clean up emissions as well as reduce the CO_2 emissions to nearly zero. However, it would still have all the impacts of coal mining and transport.

"Safe nuclear" would be based on a design that would significantly reduce the chance of a core meltdown incident whether accidental or the result of a terrorist act, and the design would be decoupled from weapon systems. If needed, this new type of nuclear plant would have to be built in a way to reduce construction time and costs compared to those built 25 years ago to merit consideration. Finally, the long-term waste storage "issue" must be resolved and the National Regulatory Commission's role of the nuclear industry watchdog needs to be restored.

Each of the nine carbonless options is evaluated, and the GtC/year reduction from each option is calculated for the base year of 2030. The total reductions from all nine options are combined and compared to the desired 1.2 GtC/year needed.

4.1   ENERGY EFFICIENCY

Wherever energy is used in our society, there are opportunities for energy efficiency. It is a diffuse and omnipresent resource. Vehicles, buildings, and industry are the three areas that are investigated where substantial reduction in carbon emissions is possible by introducing efficiency. It should be noted that efficiency is not the same as conservation. Conservation entails doing without energy services through simplifying lifestyle, frugal behavior, or deprivation. Efficiency entails doing the same with less energy using design or technology. This study limits itself to only considering energy efficiency improvements in numerical estimates. There certainly are opportunities to use significantly less energy if conservation through lifestyle changes, and this is considered in the policy chapter (Chapter 6).

Efficiency can be considered to be a "supply" option. That is, introducing efficiency improvements is equivalent to generating more supply because the energy services that are delivered are the same but with less energy used. Whether you double the automobile efficiency or increase the petroleum used in an inefficient vehicle, the total miles driven can be the same. When you focus on the bottom line, in this case miles driven, these two options can have equivalent energy services. However, you will notice that there are vast differences in the amount of petroleum used and the overall impacts of one compared to the other. The gathering, transporting, and of burning of twice as much petroleum in a vehicle has vastly greater carbon impacts compared to the impacts of developing a more efficient vehicle. If the increase in cost of the more efficient car is modest and this extra cost is paid for over several years of lower cost operation of the vehicle, then this is a cost-effective alternative to using more petroleum. This characteristic of cost-effective efficiency improvements applies across all sectors of the economy. Efficiency is an energy "source" is to be mined as one would mine coal, oil, or uranium. The opportunities and challenges are different, but the outcome is often quite favorable compared to the extraction industry.

The United States made a substantial attempt to increase energy efficiency in the mid-1970s as a response to oil price increasing nearly a factor of three in 1973 and then an additional factor of two in 1978. The results of this effort can be observed by examining the U.S. energy intensity (the amount of energy used for each dollar of gross domestic product measured in Btu/GDP$). By tracking energy intensity, you can see the overall response of the United States to the oil shocks of the 1970s.

In 1975, the U.S. energy intensity was 17,000 Btu/GDP$ (2000$). Yet this energy intensity reduced to about 10,000 Btu/$ by the year 2000, which was a 40% reduction in energy use per economic unit of production. This significant reducing in energy intensity allowed the United States to function at a total primary energy consumption of 99 quads instead of the expected 165 quads if the energy intensity had stayed on the same long-term track as it was in 1975 (a *quad* is a quadrillion Btu and equals 10 raised to the 15th power Btu/year, and a Btu is a unit of heat equal to raising a pound of water 1°F). In a very real sense, the United States saved 66 quads over those 25 years. This was due to a number of factors including energy efficiency, structural shifts in the economy, price-induced substitutions, and conservation. If energy efficiency was responsible for half of this savings (a conservative estimate), one could say that efficiency increased energy "production" by 33 quads from 1975 to the year 2000. This represents 33% of the total U.S. energy use in the year 2000 [29].

As impressive as this is, when the United States is compared to other industrialized societies, one notes that the United States uses considerably more energy in various sectors. Our transportation and building sectors are about half as efficient as Europe for example. As shown in Figure 4.2, the 40 top economies are considered and GDP is plotted versus the inverse of energy intensity (GDP/MBtu). As you can see, the United States is in the "energy-inefficient" range and better than only one (Canada) of the G8 countries and better than only 2 of the 18 industrialized countries shown. You can also see that the United States is only half as efficient as countries such as Germany,

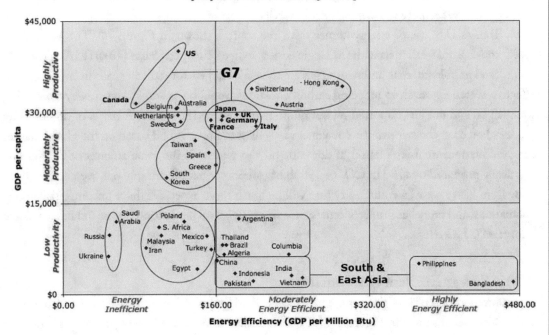

FIGURE 4.2: GDP versus energy efficiency: top 40 economies by GDP. From Wikipedia, http://en.wikipedia.org/wiki/Energy_intensity.

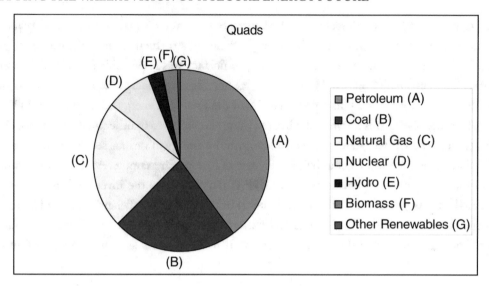

FIGURE 4.3: Primary energy distribution by source, quads. EIA.

France, Italy, the UK, and Japan. There seems to be ample opportunities to improve our utilization of energy and yet achieve the desired energy service.

Today's U.S. use of energy comes from the sources shown in Figure 4.3. The lion's share (86%) comes from fossil fuels with the single largest source being petroleum (40%) (EIA of the U.S. DOE), and petroleum, coal, and natural gas are the major targets for reducing carbon emissions. An effective overall approach to reducing carbon emissions would be to simply move away from using so much oil and not just imported oil but oil no matter where the source. A barrel of domestically produced oil costs the same to the consumer as a gallon of imported oil based on the global market economic system we use. A barrel of domestic oil also generates the same amount of CO_2 as domestically produced oil (700 lb CO_2 or 190 lb of carbon per barrel). Simply put, we need to move away from heavy use of a commodity that will be much more expensive, have uncontrollable price fluctuations, and propel us into economic chaos when global oil production peaks, and has a massive impact on GHG release.

4.1.1   Energy Efficiency in Vehicles

Transportation uses 68% of the petroleum in the United States and generates 33% of the total U.S. CO_2 emissions as shown in Figure 4.4 [30]. This makes vehicles a prime opportunity for dealing with both the peaking of global oil production and reducing carbon emissions. It is interesting to note that the total national primary energy use in 1973 was about the same as the amount used in the 1986, although the U.S. economy grew by over a factor of 3 (http://www.whitehouse.gov/omb/ budget/fy2005/pdf/hist.pdf). This was due for the most part to the vehicle efficiency [Corporate Average Fuel Economy (CAFÉ) fleet vehicle standards] and other measures enacted in 1975, and this measure doubled the efficiency of the U.S. autos from an average of 13.8 mpg to about 27.5 mpg by the mid-1980s. [It should be noted that the 1903 Model T operated at 25 mpg [31].] This national strategy of dramatically increasing vehicle energy efficiency worked before when instituted in the 1970s. Can it work again? Can the CAFÉ fleet vehicle standards be adjusted to significantly reduce petroleum use and yet deliver the desired miles driven?

In the past two decades, the average weight of the American car increased 846 lb and the horsepower doubled [32]. The number of light trucks sold in the United States exceeded passenger vehicles in 2001 and has remained greater until early 2008. Yet, because of auto industry technology improvements, we achieved about the same fleet average mileage. An amazing accomplishment! Without government leadership to guide us into a different future, the auto industry created and catered to the American car public's near-term whims and gave them fatter cars and a lot of horsepower. So clearly, we could have maintained the reasonable 1987 vehicle weight (3220 lb) and power (125 hp) and increase auto mileage by about 30% to about 35 mpg today.

This opportunity is still there. One policy approach is to raise the current fleet standard by 30% (36 mpg) for the 2015 cars and then 15% every 3 years until the standard is double today's standard to 55 mpg by 2025. The final fleet efficiency would depend on things like when hybrid-

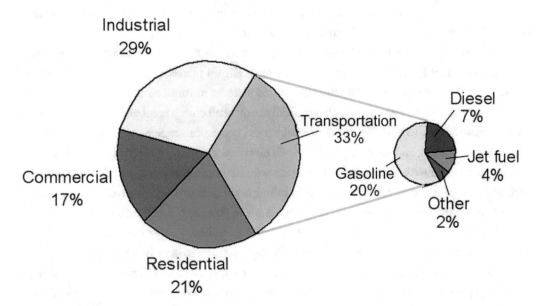

FIGURE 4.4: Carbon emissions from major energy-use sectors. From Ref. [33].

electric cars are more widespread, when diesels will be clean enough to be a common engine option, and when the hydrogen car is a reality and its actual technical characteristics. Hybrid-electric technology increases efficiency by 25% to 50%, and diesels would increase efficiency by one-third and a fuel cell by an additional one-third.

For example, if diesels were improved so that they do not degrade air quality, and were used in a diesel-hybrid electric car such as today's Prius, a mid-sized sedan could achieve a real mileage of 60 mpg instead of its current 45 mpg. This is existing off-the-shelf technology with a cleaner diesel. Also, we must abandon the fiction that SUVs including PT Cruisers are not cars and include them in the car fleet average.

Doubling the fleet mileage by 2025 would reduce our petroleum imports to less than half of today. However, there are several policy approaches to achieving these results. They will be evaluated to see whether they are cost effective, are not regressive and impact the poorer citizen adversely, are easy to administer, and the likely to actually be effective. A significant move in this direction was taken when an energy bill was signed into law in December 2007 that raised the CAFÉ standard to 35 mpg by 2020 for all new cars and light trucks. This is clearly within today's technology and could be achieved by reconfiguring the auto and truck fleet to 1987 weight and power characteristics.

Under the impetuous of upgraded 55 mpg CAFÉ standards, the vehicle efficiency would be achieved by translating the technical improvements over the last decade into mileage improvements. Beyond the initial 30%, the remaining improvement could be achieved by even lighter-weight vehicles, improved aerodynamics, and a more efficient propulsion system. This approach would also be extended to trucks and aircraft. In parallel, improvements in public transit would be pursued to give people in urban areas more of a transportation choice.

The existing engine technologies used were stoichiometric direct injection system, discrete or continuous variable valve lift system, coupled or dual cam phasers, cylinder deactivation, and a turbocharger. Transmission technologies considered were an automated manual transmission that replaces the torque converter with an electronically controlled clutch mechanism or an infinite ratio variable transmission. In addition, a vehicle's electrical system can replace mechanical components such as power steering and air conditioning, with electrical components. This increased electrification can be coupled with a high-efficiency advanced alternator. Reducing aerodynamic drag by as much as 15% and reducing energy losses from rolling resistance through low rolling-resistance tires are also important. These technologies are all available at this time in the auto industry [34].

Another opportunity is to use electricity in vehicles and I am suggesting both all-electric cars, and a plug-in hybrid-electric vehicle (PHEV). Today's use of a battery in a hybrid-electric car acts only as an energy buffer, allowing the gasoline engine to operate more efficiently. This power plant system improvement is usually accompanied by regenerative braking where the car is slowed down initially not by friction brakes but by engaging the electric motor and generating electricity that is stored in the battery for later use. The net effect increases vehicle efficiency by about 25% to 50% depending on the particulars of the design approach. For example, some designs such as the Toyota Prius is a full hybrid where the engine turns off instead of idling and the car can run solely on the battery/motor under certain conditions. Some designs are partial hybrids such as the Honda Civic where the gasoline motor is directly connected to the electric motor and does not have the flexibility of the full-hybrid designs. The results are lower first cost and lower efficiency.

The next step is to increase the battery size, design to allow a deep battery discharge, and have a plug-in capability. The car would have an all-electric range that is too short for an all-electric car but long enough to take care of the commuting distance for the average motorist. That would be an electric range of 32 miles/day for our nationally average commuter [35]. By plugging in at home at night to recharge, it would be using off-peak electricity. A liquid fuel would not be used except to drive beyond the average motorist commute distance of 32 miles/day. Thus, electricity would be used along with a fuel for the typical motorists without the range limitations and excessive cost that doomed the all-electric car.

It is interesting to note that the only item needed to adjust the infrastructure of the auto/gasoline industry to use a rechargeable hybrid-electric car would be a "dongle." This is a short (1 ft long) cable that plugs into the car's rear bumper outlet and connects to an extension cord for 120-V power. This is significantly more modest than the infrastructure requirements for hydrogen-fueled vehicle.

If your commute is greater than 32 miles, you should control the use of the 32 miles of no-emission-battery driving to be in the urban center where pollution is more of a concern.

Using electricity to power a car is cheaper than gasoline today and would certainly be cheaper than tomorrow's gasoline cost. Based on a car needing 10 kW·h to generate the equivalent propulsion of a gallon of gasoline [33], a car designed for 35 mpg could go the average commute of 32 miles at a cost of $0.92 if supplied by the national average electricity at a cost of 10 cents/kW·h. This translates into an equivalent of $1/gal cost of gasoline. If you are using off-peak electricity and with time-of-day electric rates, the cost of nighttime electricity is likely be less than 10 cent/kW·h even in the future, so the equivalent cost of gasoline would be less that $1/gal . If your car operated at closer to 50 mpg, then the equivalent cost of gasoline would be about $0.70 per gallon equivalent. The fuel cost savings per year would be $1150 for the PHEV that got 45 mpg in a world with $5/gal gasoline compared to a nonpluggable car that also has 50 mpg. The saving per year would increase to $2500 when compared to a car that has 25 mpg. There would be a strong economic incentive for a car owner to purchase a PHEV based on lower operational costs.

However, the motorist would probably need government incentives, such as tax rebates to overcome the higher initial costs of about $4000 [36] for this extended electric range hybrid-electric car. This subsidy could be paid for by a fee on the purchase price of poor (gas guzzler) mileage cars. A "fee-bate" approach would be revenue neutral and encourage consumers to "do the right thing."

This approach is being taken in Hawaii, Connecticut, and Washington, DC, where the revenue-neutral "fee-bates" tend to shift customer choice within each vehicle-size class.

Combined with the improved CAFÉ standards, and moving toward increasing biofuels, it is reasonable to have our imports in 2025 be about one quarter of today.

There is another important aspect of having a car that has a significant battery and can be plugged into standard electric outlets. When these cars get to their commuter destination, they can be plugged into the electric grid not to charge the car but to back up the electric grid. Nationally, about 6 million PHEVs would be capable of providing 150 GW (15% of total U.S. electric generation) to the grid by distributed connections throughout load centers. This is about 3% of the total number of passenger cars in the United States. This would be the electric grid's "spinning" reserve of the future, and even the peak or emergency power source if the local utility grid has evolved to a smart grid (see end note on "Smart Grid"). Pluggable hybrid-electric cars plugged in during the day in a load center could replace the need for expensive and polluting spinning reserves and peakers to give the grid required reliability and stability. PHEV car owners who plug in while at work or shopping could earn significant funds for providing this service to the electric grid. This would improve their already favorable operating costs for the PHEV.

Normally, the PHEV cars would be charged at night using off-peak power. At the present time, this energy would be provided by a combination of coal, nuclear, and natural gas because they make up most of the current electric energy system. Based on today's national electric system, each mile driven on electricity instead of gasoline would reduce the CO_2 emissions by 42%. As shown in Figure 4.5, in some states that use little or no fossil fuels to generate electricity such as the Pacific Northwest, the reduction of CO_2 from electric cars would be about 90%. Other states that rely heavily on fossil fuels such as those in a wide swat between Nevada and West Virginia would reduce the CO_2 only by about 20%. North Dakota would reduce CO_2 by zero percent because of its total reliance on fossil fuels for electricity. See Figure 4.5 for the CO_2 reduction based on in-state generation.

If the PHEVs were charged only by renewable electricity, the CO_2 released would essentially be zero. As the individual states move from fossil-based electricity generation to renewables, the amount of CO_2 released for each mile driven on electricity would decrease from 42% today to 100 % as the fraction of renewables powering the grid increased.

Some of the renewables are not baseload power systems. As large amounts of these non-baseload renewables (PV, wind, sun-only concentration solar power plants) become large (greater than 20% of the grid), then some form of energy storage would be needed to meet the needs of the grid. Wind power is greatest when the sun is weakest and tends to be a larger resource during the evening. This characteristic is well suited to off-peak charging of PHEV. So, rather than limiting wind to 20% to 30%, which would apply to today's electric grid, it may be possible for wind to

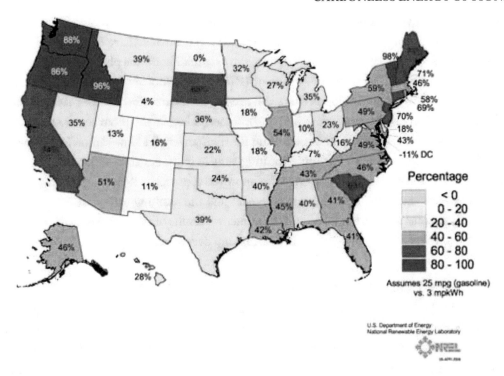

FIGURE 4.5: Carbon reduction from electric vehicles and PHEV based on current electricity generation by state. From Ref. [33].

power tomorrow's PHEV evening load and have the amount of wind on the grid to be greater than 30% without storage issues. By bringing amounts of wind greater than 30% of the grid to balance the increasing PHEV nighttime electric load, the carbon displacement of petroleum by electricity would increase in the transportation sector.

The consideration of PHEV electric storage as an integral part of the electric system can answer many questions about the ultimate amount of sun and wind-dependent renewables that can satisfactorily be used in a modern society. The battery storage in PHEVs is already paid for and will reap economic returns to the vehicle owner because it very cost-effectively displaces expensive petroleum. Because this potentially huge amount of electric storage can be made available to the electric grid via a smart grid, it can be used to stabilize the grid that has high percentage of variable renewables. This possibility requires a macro-analysis of electric grids to understand the dynamics and opportunities this by-product of the transportation sector.

Finally, hydrogen could start being used as an energy carrier (much like electricity today) in several decades with successful development of commercial automotive fuel cells and supporting hydrogen infrastructure needed for hydrogen distribution. This could become the fuel of choice for vehicles in the longer-term future. This energy carrier would be mated to more efficient fuel cell "engines," which generate electricity on board the car. Hydrogen would have to be generated by the same enlarged electric generating capacity that we started building for the pluggable hybrid-electric cars. Current research, design, and development (RD&D) support for fuel cells and support for hydrogen fuel infrastructure development are still needed for the next few decades. As important as the hydrogen RD&D is, it is even more vital to continued enlargement of our renewable electricity generation to provide the equivalent of petroleum for transportation at low carbon emissions.

There are shortcomings to simply allocating vehicle efficiency improvements via improved CAFÉ standards and noting the fuel savings. An improved approach is to develop the cost of energy savings versus the amount of efficiency improvement and the resulting carbon displacement. This economically rational approach is taken and the numerical results of this methodology are shown in Section 4.1.4, where overall energy efficiency savings are developed.

4.1.2   Energy Efficiency in Buildings

Energy use in the building sector including residential, commercial, and industrial buildings, accounts for about 43% of the U.S. carbon emissions as shown in Figure 4.6. Compared to other developed countries, America is relatively energy-inefficient in many sectors of the economy. As a result, there are many other opportunities that exist to reduce our use of energy and still maintain the same quality of life. Because of its size and relative inefficiency, the building sector is a prime area for reducing carbon.

Although many studies suggest that significant improvement in the energy efficiency of buildings appear to be cost-effective, they are not likely to occur without extensive policy changes [37, 38].

A few examples where the market does not capture cost-effective energy savings opportunities are where builders construct a home they do not live in and where first costs are valued excessively. Reduced operating costs are a secondary concern to a builder compared to few dollars more a square foot in construction costs. Also, rental buildings energy-efficient options accrue to the renter not to the builder/owner. It is very common for housing to be built that is not designed to be cost effective over the life of the building. This is even true in the commercial sector, although many studies show that worker productivity and comfort are improved by designs that reduce operating costs.

Half of today's building stock will exist in 2050, so near-term policy interventions to significantly reduce GHG emissions must quickly target this market segment with special attention to the retrofit market.

California has developed a successful model that could be used for the rest of the country. Since the mid-1970s, California has maintained its electricity use per capita at a constant value,

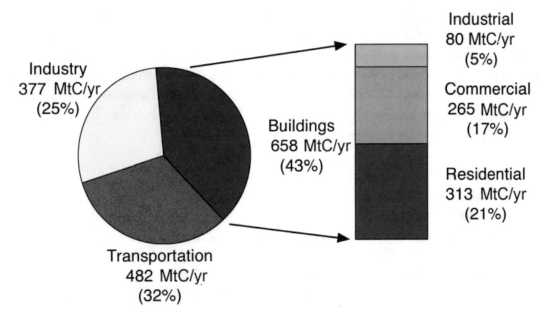

FIGURE 4.6: Carbon emissions from buildings in all end-use sectors. From Ref. [40].

although the economy has grown by several folds since then and many more electronic devices were introduced into the home, as well as a growth of 50% in the size of the average new single-family home in the United States (1500–2266 ft²). The square feet of new housing per person has doubled from 483 to 873 [40]. Thirty years ago, California electricity use per person was at the average of all the U.S. states at 6700 kW·h/person per year. Remarkably, this is the same amount used today. In the mean time, the average U.S. electricity consumption per capita has grown at 2.3%/year to13,000 kW·h/year. California has moved from average to nearly the best of all 50 states in electricity use per person.

A study found that each Californian produces less than half the GHG emissions as his or her fellow American. This is due in large part to state policies encouraging the use of natural gas and renewable resources over coal, as well as the aggressive promotion of energy efficiency. The state's per capita emissions have dropped nearly one-third since 1975, whereas the nation's per capita emissions have stayed flat.

The study notes that each Californian typically saved about $1000 per year between 1975 and 1995 just through efficiency standards for buildings and appliances. Energy efficiency has helped the economy grow an extra 3%—a $31 billion gain—compared to business as usual. The job growth created by the energy efficiency industry will generate an $8 billion payroll over the next 12 years [41].

The approach California used is a three-legged stool made up of a building efficiency code for new construction (Title 24), the CA Building Energy Efficiency Standards mandatory minimum efficient appliances, and a wide range of public utility implemented efficiency measures. This certainly could be adopted by the other 49 states and achieve similar results. During these decades, California has saved about $30 billion in energy bills while achieving this reduction in energy use, not counting the extra inflation of energy costs at the higher consumption rate.

Six plus one market transformation policies are recommended [39]:

- Improved building codes for new construction
- Improved appliance and equipment efficiency standards
- Utility-based financial incentive programs
- Low-income weatherization assistance
- The Energy Star program
- The Federal Energy Management Program
- Federal R&D support for emerging energy efficient and renewable technologies

When the buildings sector is analyzed applying these six policies, there is a reduction of 8 quads (10 to the 15th Btu/year) of energy use by 2025. R&D advances are postulated that include solid-state lighting, and a smaller contribution from advanced geothermal heat pumps, integrated equipment, more efficient operations, and advanced roofs. All together these R&D advances could contribute 4 quads by 2025. These measures are shown in Figure 4.7.

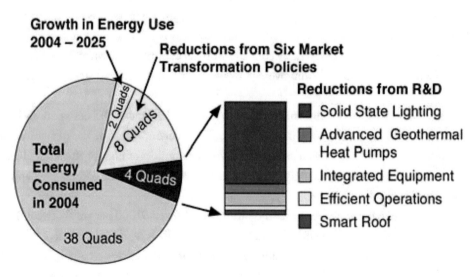

FIGURE 4.7: Building sector energy use in 2025. From Ref. [39].

4.1.2.1 Residential and Small Business Buildings. In residential homes and small businesses, the largest share of energy use serves space heating and cooling loads driven primarily by climatic conditions. There are two major opportunities for energy reduction in these buildings and it is to reduce the heat flow through the building envelope and improve the efficiency of all energy-consuming equipment. In the long run, integrated building systems have the potential of achieving net zero input of energy on an annual basis from external sources. This can be achieved by combining the energy reduction through the skin and in equipment selections, and the smaller amount of additional energy savings would come from solar hot water heaters, solar electric (PV) systems, and other on-site renewable energy sources.

Building envelop can be improved by using structural insulated panels, roof radiation barriers, roof venting, and new pigmented roof coatings that reflect summertime heat. Wall systems eliminate heat shunts of two-by framing by using structural insulated panels and insulated concrete forms. Windows available today can be six times more energy efficient than low-quality windows. There is even some electrochromic window coating under development that offers dynamic control of spectral properties of windows. They can block summer heat and light and transmit energy into the building during the winter.

Moisture inside a wall in different seasons is now better understood, and we no longer need to waste heat to avoid unwanted condensation. Thermal storage in the structure of a building can reduce energy use by buffering the temperature swings, and this can be done with lightweight materials. Even simple things such as tree planting and considering the energy involved in building demolition can make a drastic difference in the life-cycle energy use in a building. Computer tools are now available to evaluate this life-cycle energy use and the resulting carbon emission to guide building design.

Energy-consuming equipment is the other major opportunity to save energy in residential and small businesses and includes water heating, lighting, and heating–ventilation–air conditioning (HVAC) equipment. Techniques such as smarter control systems control humidity to allow higher temperatures to be comfortable, reduce air duct losses, and have variable speed air handlers. Ground-coupled heat pumps can reduce whole-house energy use and peak demand by 30% to 40%, respectively. Something as simple as matching the HVAC system size to the building heating and cooling loads have multiple benefits over "rule-of-thumb" approaches now common in the HVAC industry. Heater efficiencies have gone from the low 80s in percentage to some equipment with the remarkable level of 95% efficiency.

Water heating represents 13% of the total energy use in homes, and five technologies are now available to reduce this energy use. These are heat pump water heaters, water heating dehumidifiers, using waste heat to heat water, solar water heaters, and tankless water heaters. Finally, improving the design of the hot water distribution system is a way to reduce water heating energy.

House-based PV electric-generating systems are currently available to provide some or most of the home electricity needs. Grid-connected systems are attractive because they provide excess electricity during the day when needed typically to partially offset commercial air condition loads, and the grid can return this electricity during the evening when needed by the home. Usually, the meter runs forward and backward over a daily cycle, and electricity is sold and bought at retail price (details depend on current state ordinances). These PV panels can be added to the roof or be integrated into the skin of the building replacing traditional construction materials.

Cost reductions in PV systems have been steady for 30 years, and each doubling of production has reduced system costs by 20%. At the current rate of growth of 35% per year, a double occurs every 2.5 years. Thus, the costs are 50% lower every 7.5 years. Today's cost of about $8/w, which equals about 40 cents/kW·h (without any subsidies) and 25 cents/kW·h with California and federal subsidies, would come down to 24 cents/kW·h in 2015 without subsidies. This residential system would be more competitive with retail prices in 2015 in California. Small utility commercial systems are much larger, and systems can potentially be installed today near $4/w, which is a factor of 2 lower than individual residential systems. Without subsidies of any kind, the energy cost in southern California would be about 30 cents/kW·h. With the current federal and California subsidies, it would cost about 10 cents/kW·h, which is very attractive at today's retail prices in southern California. These numbers vary somewhat by region of the country and depend primarily on the local retail cost of electricity and the total solar insolation in that area.

Retrofit opportunities exist for efficient HVAC equipment and refrigerator replacement. Retrofitting solar water heating is fairly straightforward. More extensive use of high-efficiency fluorescent lighting in homes and small businesses could be expedited by either rebates on bulb purchases or a tax on incandescent bulbs so that its first cost is the same as that for a fluorescent.

The improvements suggested would add about 2% of the building cost when the downsizing of the HVAC equipment is considered. The payback period is about 3 years for most energy efficiency options, and the energy saving lasts over the entire lifetime of the house.

4.1.2.2 Large Commercial and Industrial Buildings. In large commercial buildings and industrial buildings, the energy use is less climatically sensitive because of relatively more building core versus perimeter, greater lighting and other internal loads, and diversity of type of business and occupancy. As a result, efficient lighting and distributed energy technologies hold great promise. Many of the technologies described in the previous section in the discussion on homes and small businesses apply equally well here. Current research holds the promise for improved wireless sensor technologies, along with advances in control technology such as neural networks and adaptive controls. These technologies are a far cry from today's simple thermostats and timers. This will allow true optimization of building energy services and reduce energy consumption and increase comfort of building occupants.

Lighting will benefit from incremental improvements over the next 20 years and will primarily be in two areas, which are hybrid solar lighting and solid-state lighting. There are also some further improvements expected in fluorescent lighting. Hybrid solar lighting uses roof-mounted solar tracking dish solar collectors that send visible portion of solar energy into light-conducting optical cables. These cables bring light to the top two floors of commercial and industrial buildings and are expected to supply light to two thirds of the floor space in the United States. This solar light is blended with variable-controlled fluorescent lighting for the hybrid application. Also, brighter LEDs are expected to further increase lighting efficiency.

A shift away from central power generation is expected toward on-site distributed energy systems that generate combined heat and electricity from the same fuel. This can double the efficiency compared to separately producing the heat and electricity. By having distributed generation throughout the grid, there are other savings in reduced distribution and transmission line losses, enhanced power quality and reliability, more end-user control, fuel flexibility, deferral of investments, and siting controversies associated with power generation and transmission.

4.1.2.3 Combining Building Sector Measures. The six market transformation policies are recommended:

- Improved Building Codes for New Construction. The greatest opportunity to make a building more efficient over its long lifetime is to make it so during construction based on good design. By requiring at least a minimum level of energy efficiency, building codes can capture this opportunity. The model provided by Title 24 in California can be adopted by each state and be stimulated by the national government. A complementary approach is fostered by Architecture 2030 started by Ed Mazria with their 2030 Blueprint at http://www.architecture2030.org/2030_challenge/index.html.
- Improved Appliance and Equipment Efficiency Standards. These standards set minimum efficiencies to be met by all regulated products sold. This eliminates the least efficient products from the market.
- Utility-based Financial Incentive Programs. When it became clear that education and information alone were not enough in the early 80s, these utility-based financial incentive programs were instituted. These programs should be continued for the duration.
- Low-income Weatherization Assistance. Residences occupied by low-income citizens tend

to be among the least energy efficient. The DOE and state weatherization assistance programs have provided energy conservation services to low-income Americans and should be continued.

- The Energy Star Program. Introduced by Environmental Protection Agency in 1992, this program fills the information gap that hinders the introduction and use of energy-efficient products and practices. The market-based approach has four parts: using the Energy Star label to clearly identify energy-efficient products, practices, and new buildings; helping decision makers by providing energy performance assessment tools and guidelines for efficiency improvements; helping retail and service companies to easily offer energy-efficient products and services; and partnering with other energy efficiency programs to leverage national resources and maximize impacts. This program should continue.

- The Federal Energy Management Program. Started in 1973, this program seeks to reduce the cost and environmental impacts of the federal government by advancing energy efficiency, water conservation, and the use of renewable energy. Specialized tools and other assistance are also proved to help agencies meet their goals. This program should be continued.

TABLE 4.1: Building sector energy savings and carbon emission reductions by 2030 from selected policies

	ESTIMATED PRIMARY ENERGY SAVING (QUAD)	ESTIMATED ANNUAL CARBON SAVINGS (MTC)
Building codes	1.5	3.4
Appliance and equipment efficiency standards	7.3	122
Utility-based financial incentive programs	0.9	14
Low-income weatherization assistance	0.1	1
Energy Star Program	1.5	30
Federal Energy Management Program	0.1	2
Federal R&D funding	4.3	72
Total	16	275

In addition to these market transformation measures, there is the additional policy of:

- Federal R&D support for emerging energy-efficient and renewable technologies. Opportunities for low-GHG energy future depend critically on new and emerging technologies. Federal R&D support in this highly fragmented and competitive sector is essential to keep the pipeline of new technologies full. R&D investment should increase to be commensurate with the opportunities for this energy-efficient form of supply.

The summary of the expected impacts of these seven measures projected to 2030 is shown in Table 4.1 [42].

4.1.3 Energy Efficiency in the Industrial Sector

The industrial sector accounts for 29% of the U.S. CO_2 emissions when buildings are included (25% without). Strategies for improving energy efficiency in the industry include efficient motors and variable drive systems, reduced piping and pumping losses, heat recovery, cogeneration, and industry-specific improvements in processes such as electrolysis. After a lull in energy efficiency activity during the late 1990s, due in part of low oil prices and the focus on restructuring in the utility sector, many initiatives have begun recently in industry and at the state level. These include a revival of utility efficiency programs such as successful experiment in Vermont with a new type of "efficiency utility" that is dedicated solely to capturing savings from energy efficiency investments.

A useful measure for a particular energy saving options is the cost of saved energy (CSE). This is the levelized net cost of the efficiency improvement divided by the annual savings of energy. This can then be compared to the cost of supply options. Each industrial operation would generate the CSE versus the amount of energy saved with options prioritized from lowest to highest cost.

The rational approach would be to choose all the actions that can be done below an upper limit of cost such as the lowest cost supply option. A minimum rate of return could also be used to decide which specific improvements to implement. Often, CSEs at one-third the cost of supplied energy with rates of return of 33% and 2.5-year payback are rejected. This apparent and very common distortion in the market for energy and energy services is one of the main reasons for policy mechanisms and utility investments to encourage efficient technology.

4.1.4 Energy Efficiency Savings

CSE-versus-cost charts look just like an energy supply-versus-cost curve and are referred to as "supply curves of saved energy." By combining the cost of energy savings analysis across all energy use, it is possible to estimate the total magnitude of this form of "energy supply." Estimates are made by reviewing limited utility data and generating a national energy saved cost curve [29].

4.1.4.1 Electric Utility Cost of Energy Savings. The electricity sector was based on work done by the five national laboratory working groups-based projects to 2020 [43]. The total technical–economic efficiency potential of the electric sector is 1500 terawatt-hours per year (TW·h/year). This is also 280 GW of capacity at an average CSE of about 2.2 cents/kW·h. This average cost is about half the cost of the least expensive supply option, which is a new pulverized coal plant. Most of the cost of coal is not included in the price because we all share the common burden of air pollution, land disruption, mercury poisoning, water way spoilage, and depleting a limited resource. Typical studies of the annual impacts of coal estimate that 24,000 people die of respiratory disease, with 400,000 acute asthma episodes (see Chapter 5).

This best estimate is reduced by 35% to be conservative and to reflect the share of total potential that is achievable given market and behavioral constraints. Thus, 980 TW·h/year (24% of U.S. total) and 180 GW (17% of total install power) are considered to be what is realistically achievable by 2020. In addition, the cost was increased by 0.6 cents/kW·h (27%) to account for implementation costs.

These data are plotted so that the average cost is 2.8 cents/kW·h, and Figure 4.8 shows the CSE in the year 2025 versus the total energy displaced. This results in a projection of about 1000 TW·h/year at a total cost saved of about 45 $/MW·h, which is 4.5 cents/kW·h and comparable to a new pulverized coal plant. Most of the energy efficiency cost savings come from industrial motor drives and air conditioning, lighting, and appliances in buildings.

Acceptable "clean" new coal plants are projected to cost close to 10 cents/kW·h, which is twice the cost of the saved energy by energy efficiency improvements.

Most of recently built power plants are very efficient combined cycle natural gas power plants which are relatively clean and low-CO_2 emitters. However, the national supply and price of natural gas is unstable compared to the market demand, and residential gas prices have fluctuated by a factor of two from 7.5 to 16 $/MBtu over the last 3 years (EIA). The industrial/utility price of gas varied from 6 to 11.5 $/MBtu, and the resulting electricity price from these advanced gas plants varied from 6.4 to 11.1 cents/kW·h. Limiting the upper price of energy saved to 4.5 cents/kW·h is conservative because prices higher than this are expected over the next few decades for new, conventional power plants.

4.1.4.2 Natural Gas Cost of Energy Savings. A similar projection of the year 2025 savings and costs for natural gas (CH_4) efficiency is made and the results are shown below [44]. To keep this

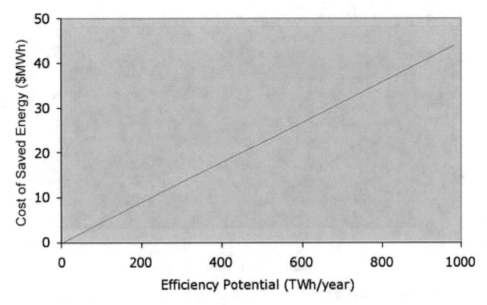

FIGURE 4.8: Electric efficiency cost curve (2025). From Ref. [29].

analysis on common footing, the same assumptions are used to de-rate the results from the electricity cost savings. That is, only 65% of the economic potential is used and an implementation cost of $0.6/Mbtu is added to the source analysis. Up to 8 TCF/year (trillion cubic feet of CH_4/year) are saved at a cost of $2.30/MBtu (1000 ft^3 of natural gas equals 1.03 MBtu) (Figure 4.9). Thus, 8 TCF/year equals 8.25 quads/year of energy saved at less than $2.50/MBtu. Most of the gas savings come from process heat and chemical feedstock used in industrial processes, and space and water heating in commercial and residential buildings.

The longer-term cost increase in natural gas is about 16% per year (EIA). This projection of about 8 quads/year by 2025 where the average cost of the saved gas is under $2.30/MBtu is a very conservative estimate if gas consumers used economic rationality to make decisions.

4.1.4.3 Petroleum Cost of Energy Savings. In a similar vane, the petroleum efficiency savings and costs are evaluated looking at transportation and industrial feedstocks. The primary mechanism considered for petroleum efficiency is not from the use of ethanol or new propulsion systems such as hybrid-electric, pluggable hybrid-electric, or fuel cells. The keys to improving energy efficiency in the oil sector rest primarily on translating the vehicle improvement over the last 20 years into better mpg rather than increased horsepower and weight. In addition, future improvements will come from lightweight materials in cars and aircraft, and advanced aerodynamics in trucks. The resulting

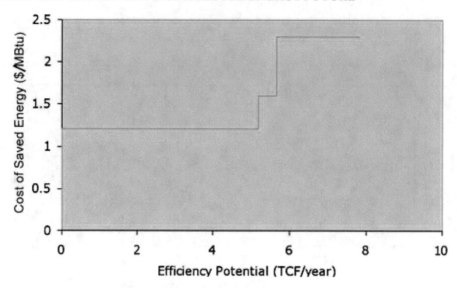

FIGURE 4.9: Natural gas efficiency cost curve (2025). From Ref. [29].

CSE curve for petroleum in the year 2025 is shown below and it includes similar assumption on implementation cost and de-rating potential.

These CSEs for petroleum are based on today's technology costs, although this projection is to the year 2025. This is assumed to allow for the time to implement efficiency programs and the time to turnover and replace capital stock. Future technology costs are expected to be lower and, to that extent, these results are conservative. This results shown below indicated that 10 Mbbl/day can be saved at less than 29 $/bbl. This is a significant amount of oil compared to the United States currently importing about 12.4 Mbbl/day, and this amounts to 21.2 quads/year. The price of oil is in the 50–150 $/bbl range now, and the oil price has been rising at 21%/year over the last decade. With the expectation of global oil peaking soon (see Chapter 2), the price of oil will have enormous pressure to go even higher. Thus, an estimate of 21.2 quads/year (10 Mbbl/day) by 2025 at a price of less than $29/bbl compares very favorably to current oil market conditions (Figure 4.10).

4.1.4.4 Overall Energy Efficiency Savings. These three sectors, electricity, natural gas, and petroleum, are combined to obtain the overall energy saved cost curve. The amounts of fuels saved are converted into carbon emission reduction to produce the bottom line—the CSE versus the annual carbon emission reduction. This is shown in Table 4.2 projected to 2030 with the extra 5 years be-

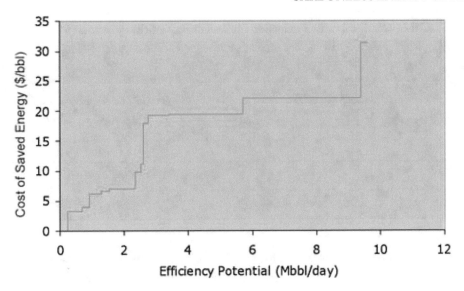

FIGURE 4.10: Petroleum efficiency cost curve (2025). From Ref. [29].

yond the results shown above to the year 2025, assumed at a 6% growth. The total resulting carbon reduction is between 635 and 740 metric tons of carbon (MtC)/year from all three sectors.

The results indicate that carbon emission reduction in the electricity sector is between 165 and 270 MtC/year as the displaced carbon varies from 160 to 260 MtC/GW·h. The sum of the carbon reductions come equally from efficiency cost savings in natural gas and petroleum for a total of 470 MtC/year. This range is the result of where the energy you are displacing comes form. If a coal power plant is displaced, the carbon displacement rate is 260 MtC/GW·h, whereas the current average for the national electric sector is 170 MtC/GW·h [metric tons of carbon per giga (10^9) watt hour].

The results for energy savings from all three sectors are shown below (Figure 4.11). This combined efficiency cost curve is the sum of the results for electricity, natural gas, and petroleum projected to 2030 and has the 35% de-rating and the inclusion of an extra cost of implementation. The middle of the range shown calculates that 688 MtC/year could be expected from all cost-effective efficiency improvements in the U.S. economy.

Based on our 2030 goal of 1200 MtC/year, this wide range of efficiency measures are about 57% of the total needed reduction in carbon emissions. These results will be combined with renewable energy options in Section 4.2.8. Energy conservation will be addressed in Section 6.8.5.

TABLE 4.2: Total carbon reduction from all energy efficiency sectors (MtC/year)

SECTOR	CARBON REDUCTION (MtC/year)	FUEL DISPLACED COST
Electricity	165–270	Less than 4 cents/kW·h
Natural gas	235	Less than 5 $/MBtu
Petroleum	235	Less than 5 $/MBtu (29 $/bbl)
Total	635–740 MtC/year	

Note: Smart Grid refers to a transformed electricity transmission and distribution network or "grid" that uses two-way communications, advanced sensors, and distributed computers to improve the efficiency, reliability and safety of power delivery and use. This evolving intelligent power distribution network includes the possibility to reduce power consumption at the client side during peak hours (Demand Side Management), facilitating grid connection of distributed generation power (with PV arrays, small wind turbines, micro hydro, or even combined heat power generators in buildings), grid energy storage for distributed generation load balancing, and improved reliability against many different component failure scenarios. It would also allow remote connection to mobile power sources such as the battery or even engines in PHEV to either back up the grid or charge on-board batteries.

4.2   RENEWABLE ENERGY OPTIONS

In addition to investigating the potential of increasing energy efficiency, renewable energy technologies are examined to see what the possibilities are for more directly meeting the goal of reducing U.S. carbon emissions by 1.2 GtC/year by 2030. To do this, 11 researchers who are expert in six technical areas were approached by the American Solar Energy Society, and were asked to establish what could be accomplished using wind, PVs, concentrating solar, biomass, biofuels, and geothermal technologies. A bottom-up approach was taken with each expert asked to come up with their best estimates of what is reasonably in their area considering resource limits, competitive economics, and existing barriers to introduction [28].

Each expert was asked to make projections based on competitive economics as well as reasonable restraints to include difficulties and barriers that may exist for each approach. Each of these six areas are examined separately and then put together to see how close this assembly of options comes to reaching the needed carbon emission goals. This effort represents both an attempt to avoid

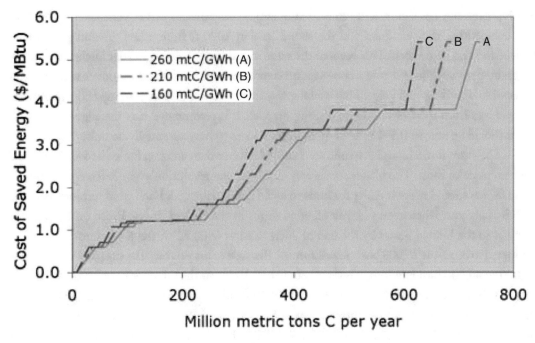

FIGURE 4.11: Total carbon reduction from all energy efficiency sectors, MtC/year. From Ref. [29].

hitting the wall of the oil production peak and that of providing the energy services to a growing society while significantly reducing the carbon dumped into the atmosphere.

4.2.1   Wind Power

Wind has made truly significant strides over the last three decades. The cost of wind electricity has gone from about 80 cents/kW·h in 1980 to about 4 cents/kW·h today—a factor of 20 times lower (EIA). The worldwide installed capacity is now 94 GW (2007) and the long-term growth annual rate is about 28%. Last year, the world installed 20 GW.

The United States currently has installed 16 GW across 34 states and is growing at 26% per year over the last decade. In 2007, 5 GW were installed at a record growth rate of 45% [45]. For a rough scale of what these numbers could mean, at the decade average rate of growth, the U.S.-installed wind capacity would be about 250 GW by 2020. This would be about 22% of total U.S. electrical power and energy.

Today, wind electric generators are about as cheap as any other source of commercial electricity. EIA projections to 2015 indicate that wind, coal, gas, and nuclear all have a projected energy cost between 5 and 6 cents/kW·h. The wind technology has evolved to its seventh generation of development and has significantly improved qualities compared to earlier versions. Besides being cost competitive, they are quiet and the sound level at 1000 ft from a line of wind generators is between 35 and 45 decibels. This is about the same sound level as sitting in your kitchen and the refrigerator goes on. You can hold a conversation standing at the base of one of today's new wind turbines without raising your voice. This is a long way from the womp-womp-womp helicopter sounds of machines from the 1980s. For good siting practice, it is recommend that the edge of new wind farms should be at least 0.5–1 mile away from residences to have essentially no noise intrusion.

The current machines have solid steel support rather than an open lattice network, and birds have no place to roost. The blades rotate very slowly at one revolution every 3–5 seconds. Both of these factors should make it easier for birds to avoid being damaged. Each wind turbine kills about 2.2 birds per year when averaged over all wind farms in the United States. Based on 12,000 wind turbines in the United States by the end of 2003 rated at 6400 MW, the total annual mortality is estimated to be about 26,000 birds for all species. Because of uncertainty, the expected range is from 10,000 to 40,000 bird fatalities. Newer machines are larger so the fatalities per turbine using older data may be misleading. Based on newer (after 1999) wind farms mostly located outside California, the fatality rate averages 2.3 per turbine per year and 3.1 per MW/year. On a similar basis, the bat mortality rate for the newer machines is 3.4 bats per turbine/year and 4.6 bats per MW/year [46].

These facts need to be placed in some context and compared to (1) other human activities that affect birds and (2) to the types of damage done by other energy systems. The buildings we all live in kill 1 to 10 birds a year on average. The road that each car uses on kills 15 to 20 birds per mile. A house cat kills 1 to 2 birds per year (see Audubon Society). All told, human activities (and house cats) kill from 100 to 1380 million birds a year (see [47], [48]).

If 20% of all our electricity in the United States came from wind farms, they would annually kill about 0.74 million birds and 1.09 million bats if we just project current national data for larger machines. Even when this large amount of wind energy is projected to some time in the future, about 0.1% of the annual harm to birds from human activities would be due to wind farms. One could conclude that bird kills from wind farms are insignificant in the general scheme of human activities.

It is important to note that the California Energy Commission's (CEC) policy is "no activity should kill birds without mitigation simply because other human activities also kill birds." This is a wise policy that should be broadly adopted. Now that a number of wind farms have been built in California and there is a better understanding of what factors contribute to higher bird kills, wind farms can now be designed to reduce the impact on birds. The CEC demands that each new wind farm be designed to mitigate bird impact based on this new analysis.

Wind turbines can be erected quickly (it takes about 1 month from faxing an order for the wind generator to generating commercial electricity at an approved site), and they tend to pump money into domestic industry and workers rather than adding to the international trade imbalance.

The visual appearance has provoked a range of responses from "they are ugly" to "they look so graceful." The issue of beauty is certainly in the eye of the beholder. The slow rotation speed seems to add to the visual attractiveness of this renewable energy source. Care should be taken when siting a wind farm to avoid bird migration routes or visually sensitive areas.

Current wind generators each can produce about 1 to 3 MW of power. They now generate about 1% of the total electricity used nationally. Projections to 2030 are for the cost to drop below 3 cents/kW·h. A host of factors have brought down the cost of wind power. The materials used have improved and turbines are now much larger and more efficient: 125 m in rotor diameter compared with 10 m in the 1970s. An important improvement is that the cost of financing wind farms also has dropped as financial markets become more comfortable with the risks involved.

The only factor that has modulated this rapid growth rate has been investor uncertainty over the government support role. The production tax credit (PTC) is being used by the U.S. government to stimulate wind power, but the support has not been applied uniformly. This support has occasionally been allowed to lapse with disastrous results causing numerous projects to stall. It would be far more efficient for this government policy to be steady and predictable over the long term. This action would be a key to future rapid growth. An example of such a policy would be for the existing PTC of 1.8 cents/kW·h to be continued until 2010 and then phased out linearly until 2030.

The national wind resource is huge. Except for favorable locations along the ridge lines in western and some eastern mountains, wind on land is primarily in 14 midwestern and western states between northern Texas and Montana [49]. There is also a sizable off-shore wind resource which is not considered in the study. It could easily generate all of our current national electricity by using about 2% of the land in these states. This power system can coexist with the farming and ranching in this vast region, which reduces the land use conflicts. In fact, it allows ranchers and farmers to multiply their cash flow and allows them to maintain their lifestyle without uncertain financial pressures. Within a wind farm, only 2% to 5% of the land is actually used for the wind farm structures and support system. Almost all of the land is still available for the existing use.

Wind is strongest in the non-sun times of the day during the summer as shown in Figure 4.12.

However, wind does a better job matching utility load in the winter as shown in Figure 4.13.

Wind use would normally be limited to 20% or 30% of the regional electric grid to avoid grid instabilities due to the relative unpredictability of wind and the diurnal and seasonal variability.

Beyond the 20% to 30% level of wind use, the conventional approach would be to store the excess evening energy in underground geologic features using compressed air or by using pumped hydro plants. Eventually, conventional thinking would plan to use hydrogen, when it becomes the

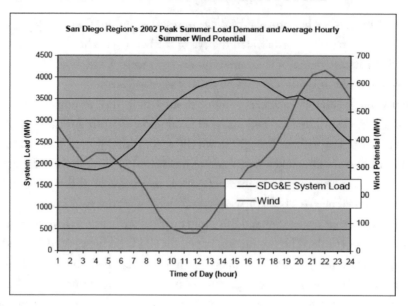

FIGURE 4.12: San Diego's 2002 peak summer load demand and average hourly summer wind potential.

energy carrier of choice, as the system storage medium, and it could be pumped into depleted underground oil and gas fields in the southwest and upper west where pipelines already exist to move the gas to market.

These storage techniques may prove to be unnecessary. One of the thrusts to moving away from oil use will be the promise of the pluggable hybrid-electric vehicle. Nighttime vehicle charging is a good match to wind power, which is primarily a nighttime energy source. This match eliminates the need to store wind for the normal electric-grid daytime loads. Wind can be grown nationally beyond the 20% to 30% limit as the chargeable hybrid-electric vehicles are introduced into the fleet especially if the vehicle charging is a load controllable by the utility in a smart grid scenario.

An interesting electric system characteristic of this development is that some of these rechargeable PHEVs can be plugged in during the day while at work. The plug-in vehicle would not be to charge the cars batteries but to generate mid-day peaking electricity for the electric gird and be a back-up stand-by for electric grid reliability purposes. If 3% of our future vehicles were PHEV, the electric-generating capability of these vehicles would be 150 GW, which is 15% today's total electric generation capacity. This would drop the need for the inefficient fossil plants now used to provide grid spinning reserves and stand-by.

With this enormous potential of wind, care must be taken to make projections in the 2030 time frame that recognized early limitations to this attractive energy source. The National Renew-

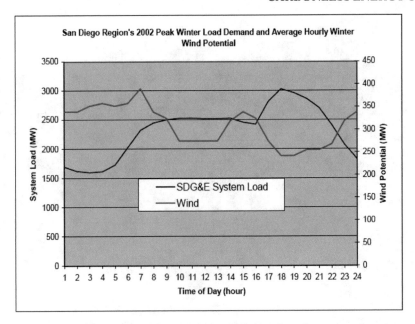

FIGURE 4.13: San Diego's 2002 peak winter load demand and average hourly summer wind potential.

able Energy Laboratory (NREL) has assembled a comprehensive geographic information system (GIS) system for renewable generation. The GIS data set for wind is based on a large-scale meteorological simulation that "re-creates" the weather, extracting wind speed information from grid points that cover the area of interest. Mapping processes are not always directly comparable among the states, but NREL has validated maps from approximately 30 states [50].

The GIS database contains wind speed at 50 m from the ground, and this creates a mismatch with currently available wind turbines with a hub height of 80 m. The hub heights will increase further in the future to 100 m or even higher. This implies that the estimates of the wind resource shown in the map below will likely understate the wind resource, perhaps significantly, resulting in supply curves that also understate the quantity of wind at various prices. Figure 4.14 shows the U.S. wind resource map. This is a relatively broad resource and is dispersed over wide swats of the country. The only area low in wind resource is the southeast. The Wind & Hydropower Technologies Program, U.S. DOE, has developed maps for much of the United States under its Wind Powering America site, available at http://www.eere.energy.gov/windandhydro/windpoweringamerica/wind_maps.asp).

A set of wind supply curves was developed for the Western Governors' Association (WGA) study [Wind Task Force (WTF) of the Western Governors' Association's Clean and Diversified

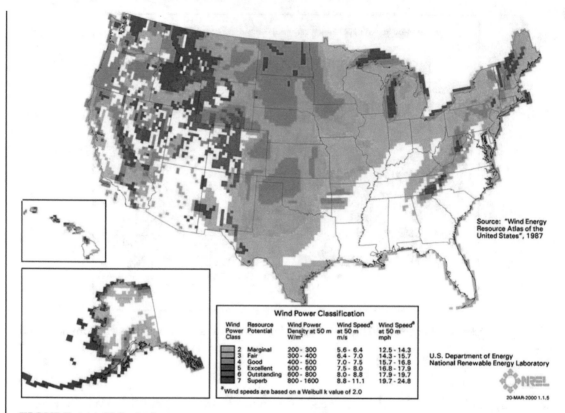

FIGURE 4.14: U.S. wind resource map.

Energy Advisory Committee (CDEAC)]. These were augmented by a range of other studies [50]. As part of this extensive analysis, a new innovative transmission tariff was evaluated that provided long-term transmission access on a conditional firm or nonfirm basis. The WTF concluded that such transmission tariffs could speed the development of wind and increase the efficiency of the transmission system. Three scenarios were generated by 2015 assuming little, midrange, and high-end transmission tariff development. The results show a range a factor of more than 5 from 10 to 55 GW and indicate the importance of treatment of transmission tariffs.

The GIS wind resource map shown above is modified to place limits on the amount of wind that can be expected. Wind development is excluded from urban areas, national parks and preserves, wilderness areas, and water. Several sets of supply curves were developed for each state and then combined for the entire country.

A key factor is wind's access to existing transmission lines. A range of assumptions were made ranging from no access to as much as 40% of the nearest transmission line is available. When no

line is available or when the amount of wind uses up the available transmission capacity, a new line is built to the nearest load center at a cost of $1000/MW-mile.

The resulting wind supply curve is shown in Figure 4.15 where the cost of wind energy is portrayed versus the amount of wind capacity. At a distance of 100 miles from an existing transmission line, the additional cost of transmission would be about 0.5 cent/kW·h (5 $/MW·h) at the assumed cost of $1000/MW-mile. This is a 10% cost increase for wind not near existing transmission lines. So, assumptions about transmission issues can have a strong bearing on the resulting projection of wind power. If delivered cost is limited to 6.5 cents/kW·h, then the wind-resulting wind capacity would be from 30 to 255 GW depending on transmission line availability as shown below [50].

This wind supply curve is also shown in Figure 4.16.

Using DOE cost projections of from 2 to 3 cents/kW·h for future wind machines built after 2015 in class 6 wind, a projection of wind introduction is simulated by a NREL model [51]. This model characterizes the entire U.S. power grid by control area and superimposes the GIS representations shown above along with actual transmission line location. The wind cost data push against all other generation technologies with price forecasts and determine the least-cost national generation portfolio that matches demand. The key external assumption is that the PTC is renewed until 2010 and is then smoothly phased out by 2030. The results are shown in Figure 4.17, and the amount of wind is about 20% of the U.S. energy consumption served by wind.

When the amount of wind reaches 20% of the wind served market, it is caped at 20% to avoid any grid stability issues. Also, the opportunity to use system storage or evening time wind energy to power pluggable hybrid-electric vehicles is ignored in this conservative estimate.

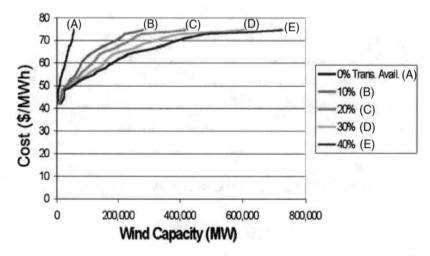

FIGURE 4.15: Supply curves for different percentage of existing transmission available to transport wind to load. (from Ref. [50])

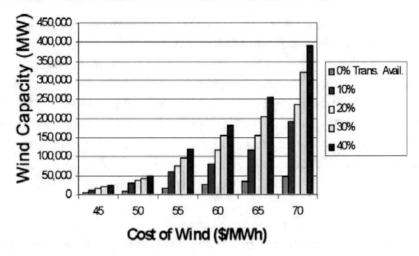

FIGURE 4.16: Impact of transmission availability on wind supply. (from Ref. [50])

These results are translated into carbon reduction potential. There is a range of results to reflect the variation in the carbon content of the displaced electricity in different parts of the country from 160 to 260 Mt/GW·h. The results shown in Figure 4.18 project between 140 and 225 MtC/year in the year 2030. The midrange is about 180 MtC/year of carbon would be saved by this renewable technology.

When the accumulated carbon reduction is calculated as shown below, the average projection is about 1400 MtC by the year 2030. This result will be summarized with all the other carbonless options in Section 4.2.8.

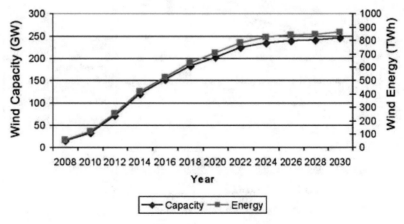

FIGURE 4.17: Results of high-penetration wind market simulation. (from Ref. [50])

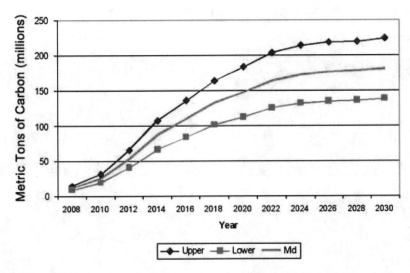

FIGURE 4.18: Range of annual carbon reduction from wind. (from Ref. [50])

4.2.2 Concentrated Solar Power

Concentrating solar power (CSP) plants can be utility-scale generators that produce electricity by using mirrors or lenses to efficiently concentrate the sun's energy to drive turbines, engines, or high-efficiency PV cells. Because these solar plants track the sun, they produce power over most of the day and can produce energy at summer peak-demand periods. There are five significant CSP technologies, and they are designated by the type of equipment that collects the sunlight. They include systems build around parabolic troughs, dish-Stirling engine systems, central receiver, compact linear Fresnel reflector, and concentrating PV systems (CPVs) as shown in Figure 4.19. The utility PV systems could be expanded to include flat plate designs that are either tracking or fixed.

Trough, Compact Linear Fresnel Reflector, and central receiver configurations include large central power blocks, usually steam turbine-generators for MW-scale output, whereas dish-Stirling and CPV systems are composed of a large number of smaller modular units of about 25 kW each that can be amassed into MW-scale plants.

Parabolic trough systems have been deployed in major commercial installations built by the Luz company in the 1980s. Nine parabolic trough plants, with a combined capacity of 354 MW, have been operating in the California Mojave Desert for two decades. These plants were built as the result of attractive federal and state policies and incentives available for renewable energy projects at the time. The parabolic trough plant was recently reinvigorated when a 64-MW steam-Rankine trough plant in Nevada that was completed in 2007. State renewable energy portfolio requirements are driving the construction of most of these CSP systems.

TYPES OF CSP TECHNOLOGY

Amonix Concentrator PV

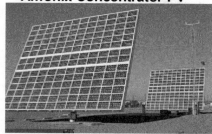

Central Receiver

Parabolic Trough

Central – Kramer Junction (CA) Solar Electric
Generating Station

SAIC and SES Solar Dish Systems
in Operation
UNLV Installation, 8/17/01

Dish-Stirling

Ausra Compact Linear Fresnel Concentrator

eSolar Heliostat Sticks

FIGURE 4.19: From http://www.ausra.com/.

A second CSP technology is just entering large commercial status and it uses the dish-Stirling module approach. A Stirling Energy System (SES) design has a 37-ft-diameter parabolic dish reflecting the sunlight on to the head of an efficient Stirling external combustion engine. This kinematic Stirling engine powers a small generator producing about 25 KW of electricity at each module. The motor generator, about the size of a motorcycle engine, is enough to provide electricity to 5 to 10 homes. The approach planned is for these identical modules to be factory mass produced to achieve low cost as much as cars are mass produced. There would be only repetitive on-site assembly without the complications and expense of site-specific large central power plant construction. Electricity would be collected via underground wires to a central substation for connection to the electric grid. A plant called Solar 1 by SES will be rated at 500 to 850 MW and is currently under a utility [Southern California Edison (SCE)] power purchase agreement contract for siting north of Los Angeles. Another large plant called Solar 2 is rated at 300 to 900 MW and will be located east of San Diego in the Imperial County desert.

A second Stirling module design approach is being taken by Infinia using a free piston Stirling engine rather than a kinematic design with crank shafts and connecting rods. This appears to be a simpler and lower-maintenance design that would reduce operation and maintenance costs. The approach is based on a 3-kW design that can be ganged into multicylinder design at 30 kW. This approach can be used for within-the-grid distributed applications at lower power levels or massed for a central power plant. At the present time, Infinia is obtaining development and precommercial financing but has not signed a Power Purchase Agreement with any utility.

A recent contract was signed between PG&E and BrightSource (Luz II) Company for a central receiver power plant rated at 100 to 500 MW. Luz II is using the DPT 550 design where 100 MW modules are used as the basic building block of the power plant. In addition, smaller commercial CPV plants exist or are planned such as the 1MW of Amonix Concentrating PV to be located in Nevada and a new commercial plant in Australia.

The compact linear Fresnel reflector is a recent entry into the CSP sweepstakes. Ausra, an Australian group who recently relocated to Palo Alto, has signed a power purchase agreement with PG&E for a 177-MW project using commodity flat mirrors that are focused on a linear pipe which heats water to steam at 545°F for use in a conventional steam turbine power plant. Although it operates at about 100° lower temperature than the parabolic trough solar field, the promise of lower cost collectors may make this approach cost competitive with other CSP approaches. Thus, all five CSP technologies are either commercially available or under contract to start construction soon.

All of these technologies have some characteristics in common, such as using direct normal solar insolation (not diffuse sunlight), which is primarily a southwest U.S. resource. Because there is little diffuse light, a daylight shadow in the southwest is so dark even on a sunny day. The parabolic trough, CLFR, and the central receiver optimum commercial size is from 100 to 500 MW but can be modularized to 1000 MW or more. The dish-Stirling and CPV plants start commercial viability around 5 MW and can be built up to 1000 MW or more.

The parabolic trough plant siting requires land slope of less than 1° while the dish-Stirling and CPV can use land up to a 5° slope. The central receiver has a slope requirement between 1 and 5 degrees. Because the parabolic trough and central receiver commercial size is larger, there is a minimum continuous land requirement from 0.8 to 4 square miles. The dish-Stirling and CPV can be sited on irregular land plots and land that can be as small as 25 acres at the 5-MW minimum size.

A remarkable feature of the parabolic trough and the central receiver plants is their ability to use thermal storage to extend their ability to produce power beyond the time of sun availability. This additional operational time each day is achieved at nearly no increase in energy cost. However, the capital cost does go up nearly linearly with hours of thermal storage. These technologies could relatively easily be operated on other fuels such as natural gas of bio-liquids for hybrid operation by the simple device of adding a boiler to generate steam at the existing central power block. These characteristics are important when larger amounts of CSP are being introduced into a grid where a larger capacity factor becomes more important. The CPV technology can only produce power in the sun-only mode of operation.

The initial dish-Stirling plants will be sun-only operation with a capacity factor of about 0.30. That is, it produces rated power about 30% of the day on average. Dish-Stirling can have hybrid operation by using a dual-fuel engine design (sun and natural gas or bio-liquids) to extend the time of operation each day and allow load following operation. Surprisingly, for a small (25kW) engine, the hybrid efficiency is about 38%, which is better than most large fossil power plants (except combined cycle gas power plants and some ultraefficient coal plants). For hybrid operation, an engine with a redesigned heater head for the dual fuel needs to be developed. The extra cost of this engine, the cavity collector cover plate to accommodate combustion, and the fitting of the hybrid fuel lines to each dish module needs to be determined. It is possible that it would be cheaper to use a gas turbine peaker for the few hours a day of extended operation.

These four technologies are currently available and one is pending. These should provide a strong stimulation to the competitive development of this renewable energy source. If any of the four of these five technologies now making the transition from advanced prototype to commercial operation falter, it will be rather straightforward to move to one of the alternatives. Studies project parabolic trough initial power plants to be in the 15–20 cents/kW·h cost range with expectations around 10 cents/kW·h by 2015 (see Figure 4.20).

This analysis developed for the U.S. DOE's Solar Energy Technologies Program multiyear program plan describes cost reductions for parabolic trough systems based on a combination of continued R&D, plant scale-up, and deployment [52]. For this analysis, deployment levels were conservatively estimated at 4 GW by 2015, based on consensus within the WGA working group.

When the central receiver technology reaches commercial operation, it is projected by some analysis to be about 25% less expensive than the trough. Similarly, the dish-Stirling power plant is expected to be about 25% less expensive than the central receiver after commercial status is established with mass-produced engines and dish components. Projections for the CPV commercial

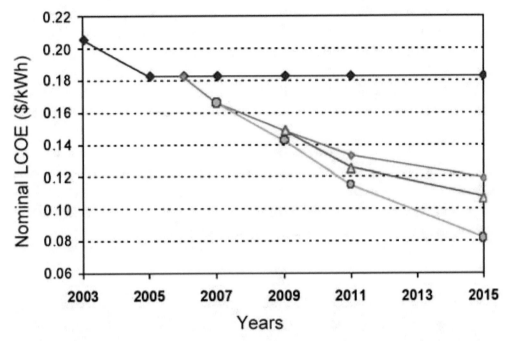

FIGURE 4.20: Projections of CSP systems energy cost versus scale-up and R&D. From DOE.

system energy cost overlays most of this range, and is projected to be 9.5 to 22 cents/kW·h [53]. These costs depend on actual operational commercial data, and only that type of data will establish the actual ranking of these CSP systems.

The amount of direct normal sunlight in the southwest is enormous and is located primarily in seven states centered on Arizona as shown in Figure 4.21 (lower left corner) in darker reds and deeper oranges. Not all the land area shown is suitable for large-scale CSP plants, and to address this issue, GIS data were applied on land type (e.g., urban, agriculture), ownership (e.g., private, state, federal), and topography. The terrain available for CSP development was conservatively estimated with a progression of land availability filters as follows:

- Lands with less than 6.75 kW·h/m² per day of average annual direct-normal resource were eliminated to identify only those areas with the highest economic potential.
- Lands with land types and ownership incompatible with commercial development were eliminated such as national parks, national preserves, wilderness areas, wildlife refuges, water, and urban areas.
- Lands with slope greater than 1% and with contiguous areas smaller than 10 km² (about 4

square miles) were eliminated to identify lands with the greatest potential for low-cost development. (This is quite conservative because land with greater slope up to 5° and parcels as small as 25 acres are suitable for some CSP technologies.)

The larger part of Figure 4.21 shows the land area remaining when all of these filters are applied. The resulting potential power generation is still enormous and is about 5% of the 1 million square mile Southwest lands [54].

Table 4.3 provides the land area and associated CSP generation capacity associated with the remaining land in this figure. This table shows, even if we consider only the high-value resources, on nearly flat, large parcel of land, that nearly 7000 GW of solar generation capacity exist in the U.S. Southwest. To place this in context, the current total U.S. electric power is about 1000 GW of generation capacity (EIA). This is a very large renewable energy resource.

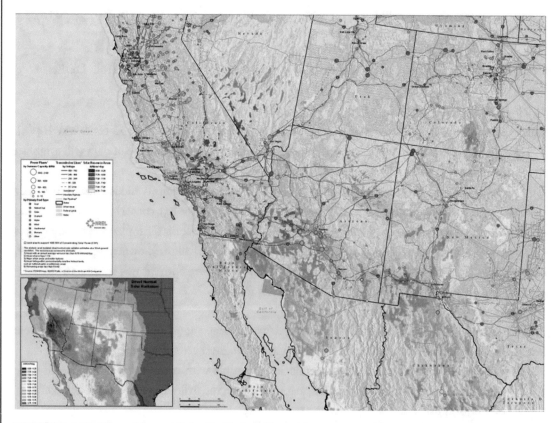

FIGURE 4.21: Direct Normal Solar Radiation, filtered by resource and use and topography. (from Ref. [54])

TABLE 4.3: Concentration solar power resource magnitude (from Ref. [54])		
STATE	AVAILABLE AREA (m²)	CAPACITY (GW)
Arizona	19,300	2,470
California	6,900	880
Colorado	2,100	270
Nevada	5,600	715
New Mexico	15,200	1,940
Texas	1,200	149
Utah	3,600	456
Total	53,900	6,880

To estimate the amount of CSP plants that could be installed by 2030, cost–supply curves developed for this analysis are based only on a 100-MW parabolic trough because, at this time, it is the only operating commercial CSP technology. The trough plant is considered to have 6 h of thermal storage and is shown for a range of solar insolation values (6.75–7.75 kW·h/m² per day). Six hours of storage would increase the annual capacity factor for the plant from about 0.29 for sun-only operation to about 0.43 and depends to some extent on the richness of the solar resource. This would allow more load following capability than sun-only operation. (The annual amount of energy generated is equal to the rated capacity times the capacity factor.)

A 20% transmission-capacity availability to the nearest load center is assumed. Where the solar resource is located adjacent to a load center, 20% of the city's demand is assumed to be available to be filled by the solar generation without the need for new transmission. The analysis assumes that when the 20% capacity is allocated, new transmission must be built to carry additional supply to the nearest load center. New transmission cost is assumed to be $1000 per MW-mile. Nominal levelized cost of energy (LCOE) is used to estimate energy cost and represents a utility financial analysis approach. The amount of CSP is shown at different levels of direct normal solar insolation (note: 10 $/MW·h = 1 cent/kW·h).

The resulting supply curves shown below are based on the current transmission grid (Figure 4.22). However, they provide a qualitative assessment of the cost of energy from future CSP plants because the curves include the cost of constructing new transmission to areas where lines do not currently exist.

To understand how cost projection would affect the rates of introduction of CSP plants, the Concentrating Solar Deployment System Model (CSDS) was used. CSDS is a multiregional, multi-time period model of capacity expansion in the electric sector of the United States and is currently being used to investigate policy issues related to the penetration of CSP energy technologies into the electric sector. The model competes parabolic troughs with thermal storage (based on fixed capital and variable operating costs over 16 annual time slices) against fossil fuel, nuclear, and wind generation technologies. Costs for nonrenewable generation are based on the EIA's Annual Energy Outlook 2005 (see Section 4.2.1 for description of unrealistically low EIS projections of conventional energy). The CSDS model was developed recently by the NREL [55].

The results shown below indicate that up to 30 GW of parabolic trough systems with thermal storage could be deployed in the Southwest by 2030 under an extension of the 30% income tax credit. Under a more aggressive scenario assuming the introduction of a $35/ton of CO_2 carbon tax, the results indicate penetration levels approaching 80 GW by 2030. Extension of this analysis to include the identical transmission constraints used earlier to develop the CSP cost supply curve

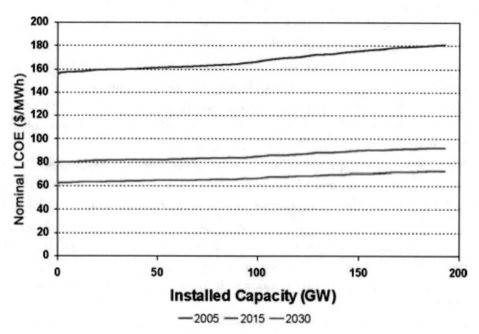

FIGURE 4.22: Transmission limited supply curve. (from Ref. [54])

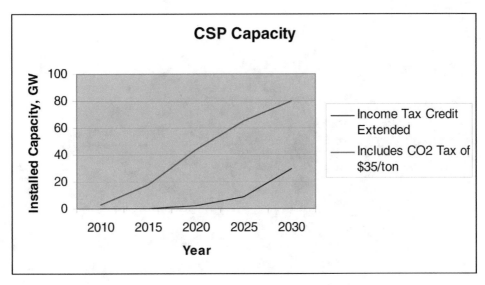

FIGURE 4.23: CSP supply curves. (from Ref. [54])

results in the supply curves shown below (Based on Nathan Blair, Mark Mehos, Walter Short, Donna Heimiller, Concentrating Solar Deployment System (CSDS)—a new model for estimating U.S. concentrating solar power (CSP) market potential. NREL/CP-620-39682, April 2006). It should be noted that the supply curves in Figure 4.23 are based on the current transmission grid. These results provide a qualitative assessment of the cost of energy from future CSP plants because the curves include the cost of constructing new transmission to areas where lines do not currently exist.

Future carbon displacements were based on the penetration and costs described by the supply curves and were estimated using a conversion factor of 210 metric tons of carbon per gigawatt-hour of displaced generation. The resulting carbon displacements would be about 25MtC/year if 30 GW are on-line in 2030. When 80 GW of solar are projected, it would replace about 65 MtC/year. This result is combined with the other renewable energy options in Section 4.2.8.

4.2.3 Photovoltaics

PV flat panels located on or near buildings can generate electricity from sunlight with no moving parts or noise. Possible locations range from roofs, south-facing walls of commercial building, to parking lot shading structures. Because this power supply is located in the electric grid load center, it reduces the cost of the power infrastructure because it does not add transmission lines and either

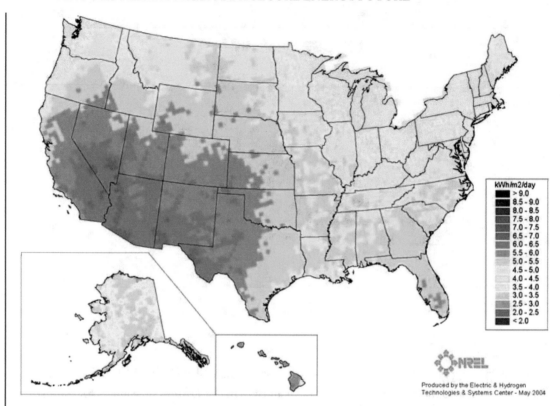

FIGURE 4.24: Total solar insolation on flat plate tilted at latitude angle.

forestalls or delays distribution system upgrades. It uses both direct normal and diffuse sunlight and as shown below; the solar resource typically varies by ±26% (4.25–6.75 kW·h/m² per day) around almost all the country. This renewable resource is available in all population centers and can be used over the entire country. The solar energy levels shown below are for a fixed, flat surface tilted to the south at the local latitude angle (Figure 4.24).

This technology emerged from aerospace applications in the 1960s and has developed commercially starting and growing in niche markets of increasing size over the last 40 years. Efficiency and cost have improved significantly over the decades with efficiency going from 4% to about 18% for current commercial systems. R&D cell efficiency is now in the range of up to 40% for multijunction cells , up to 24% for single crystal cell, and up to 20% for thin-film cells [56].

Cost has gone from about $90/w in 1968 [57] for the PV panel (main component of the system) to about to $3/w today. The panel is about half the total installed system cost for most systems.

The 2005 U.S.-installed PV power is 150 MW, which is well less than 1% of the U.S. total; and the U.S. PV capacity is about 10% of the world total of 1700 MW. However, the rate of growth of PV within the United States has been 40% per year over this decade, and the global long-term historical growth rate is 29%. Cost reductions in PV modules have been steady for 30 years, and there has been a consistent cost reduction of 20.2% (learning rate) for each doubling of production [58]. Other evaluations of the PV module's long-term production and cost reductions conclude that the learning rate is about 18% per doubling of production [59].

The balance-of-system (BOS) is made up of everything besides the PV modules and typically are the inverter (convert from DC to AC), the brackets to hold the panels, wire, installation, permits, site engineering, and profit. The PV system is made up of the panels and BOS. The BOS makes up about 50% of the total system cost today, and there is some evidence that BOS costs have fallen along with, if not faster than, module costs [60].

This means that PV system cost is decreasing by about 7% per year on average. At this historical rate of growth of 29% per year, a doubling of production occurs every 2.7 years. As a result, the installed PV costs are reducing by 50% every 10 years if the system costs followed the long-term historical trend.

Today's rooftop retail cost for a crystalline silicon PV system is still expensive for bulk electric power and costs about $7/w (commercial) to $8/w (residential) [61]. Without subsidies for a residential application, the energy cost over the 35-year life of the system would be about 31 cents/kW·h in southern California assuming an annual capacity factor of 0.2. This includes the cost of a loan by the home owner at 6% interest for the 35-year system life for all up-front costs as well as the cost of maintenance assuming that a $2000 DC/AC inverter is replaced every 10 years, and that there is tax deduction for the loan interest. For reference, wholesale electricity is usually in the 4 to 15 cents/kW·h range and low-usage gas peaker power plants cost up to 35 cents/kW·h when gas cost is high. The residential installation cost is reduced by the 30% federal tax credit. In addition, the installed cost is 10% to 15% less ($7/w) if the PV system was put on new homes in multihome development or a large commercial roof installation with some efficiencies of scale. The resulting energy cost for a commercial installation is less because of accelerated depreciation and other tax benefits. Both the residential and commercial installation may benefit from other state incentives that are either a $/w rebate or a feed-in tariff at so many cents per kilowatt-hour. Bulk purchased thin-film PV systems cost between $3.50/w and $5.50/w installed and result in about 25% to 50% lower energy cost.

The commercial economics took a positive turn recently when the California Public Utility Commission approved the first power purchase agreement between a utility and a PV system supplier for a roof-mounted system. The SCE has selected First Solar to engineer and supply the PV power plant system for a 2-MW project to be installed on the roof of a commercial building in Fontana, CA. This is the first installation in SCE's plan to install 250 MW of solar-generating capacity on large commercial rooftops throughout Southern California over the next 5 years. First Solar is using a thin-film PV technology and has achieved the lowest manufacturing cost per watt in the industry—$1.14/watt for the first quarter of 2008. The suggested installed price of this system is about $3.50/w and a stepwise improvement in PV installation cost. SCE will make a financial arrangement with the building owner and the power will be pumped directly into the local grid. This goes beyond the existing grid-tie arrangement where the building owner owns the system and can only benefit from the generated PV energy up to the energy used by the building on an annual basis.

At this time in California, the PV system cost is reduced to 17 cents/kW·h with rebates and federal 30% tax credit for a residential homeowner. This cost is about 15 cents/kW·h for a commercial installation due to the 30% federal subsidy and accelerated depreciations. Thin-film installations would be about 20% less at this time.

PV systems are now knocking on the door of the mainstream electricity market in good sun and high retail electricity cost areas. With a grid-tie arrangement, the cost of the PV energy system owned by the home or building owner is compared to retail prices (grid-tie simply runs the meter forward and backward as more electricity is generated than used or the reverse). This retail price is at least twice the cost of wholesale energy. Selling on-site electricity at retail price means that the PV systems have gained about 10 years in reaching cost competitiveness.

With the continued average 18% to 20% reduction with each doubling of production, PV will enter sunnier retail markets in the next 5 years and the entire country within about 30 years without subsidies. DOE's Solar America Initiative cost goals for a rooftop PV system are shown in the figure for residential (2–4 kW) systems to 2030. This projection of goals comes close to that if one assumes the historical growth rate of 29% and 18% learning rate at least to 2025. The DOE goals are somewhat lower that the continuation of the historical trends but then tend to flatten by 2025. The recent SCE/First Solar power purchase agreement shows improved cost performance compared to this DOE goal projection at least in the near term (Figure 4.25).

In 2015, based on the system cost reduction of 50% due to continued global production, this southern California residential retrofit cost would come down to 16 cents/kW·h in 2015 without subsidies of any kind. Similarly, the commercial installation would be about 12 cent/kW·h in 2015. These costs to generate energy on-site with a grid-tie connection to the local grid would be more than competitive with retail prices in California where there are high retail electricity costs and good sun.

These numbers would vary by region in the country proportional to the magnitude of the solar resource shown above and the relative cost of retail electricity in that region. The approximate order that some major cities will become cost competitive with local electricity costs are southern California at $5.10/w, San Francisco at $4.5/w, Los Vegas at $4.10/w, Miami at $4/w, Phoenix at $3.90/w, Boston at $3.60/w, Denver at $3.10/w, Philadelphia at $2.90/w, Atlanta at $2.80/w, Chicago at $2.30/w, Salt Lake City at $2.20/w, Fargo $1.80/w, and Seattle at $1.2/w [59]. When these system cost estimates are compared to the projection of system cost versus time, it is possible

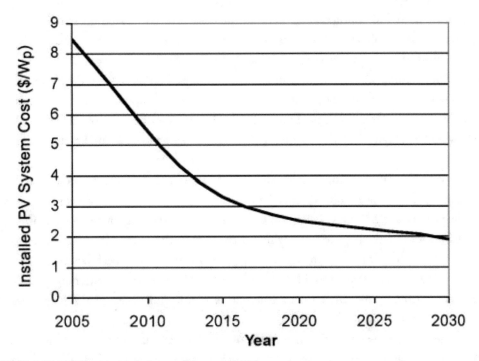

FIGURE 4.25: DOE cost goals for roof-mounted PV system.

to estimate the year when each location will be cost competitive without subsidies. In Southern California, it will be around 2012 and after 2030 in Seattle. The other cities will fall in between these two extremes.

The amount of world installed PV would be about 1000 GW in 2030 at these growth rates; and if the United States is about 10% to 20% of the world total, then the US-installed PV would be from 100 to 200 GW. The resulting energy cost of rooftop PV would start at today's cost of 31 cents/kW·h at an average location (capacity factor = 0.2) in southern California assuming the building owner borrows the up-front cost at 6% paid back over the life of the system, and can deduct interest from taxes, and replaces the inverter every 10 years. The PV energy cost would vary from 31 to 49 cents/kW·h from the location with the best solar resource to the poorest without subsidies. The estimated cost of residential electricity from solar PV would go to around 16 to 25 cents/kW·h by 2015 and 8 to 12 cents/kW·h by 2030. The 2030 rooftop PV cost would be competitive with wholesale electricity at that time. Costs for commercial scale (10–100 kW) and utility scale (1 MW or greater) are expected to be even lower. The expected price reduction should come from module cost reductions, module efficiency improvements, economies-of-scale for aggregated and larger PV markets, and improved system designs.

As with wind and concentrating solar thermal systems, PV placed on buildings is a large energy resource. Available roof area in residential and commercial buildings is about 6 billion m^2 even when most area is eliminated for shading and poor orientation related to south. This does not include agricultural and certain industrial buildings. Based on an average PV power density of about 100 w/m^2 for a silicon roof system, this potential roof area would generate about 600 GW, which is close to the current U.S. total power level. The amount of energy 600 MW of PV would generate would be about 1000 TW·h/year, which is about 28% of the total U.S. annual electricity energy [62].

At low levels of penetration, it is an advantage to a grid to have PV generated electricity during the day when there is a large commercial load. In places with summer air conditioning loads driving the peak power, PV does push again this peak load but typically provides from 50% to 70% of its rated power to reducing the peak, which occurs late in the afternoon. This is because if PV panels are tilted south for maximum annual energy, then they produce peak power occurs at noon ±2 h. This power profile peaks several hours before a late afternoon air conditioning load. In places with a winter evening peak rather than a summer peak, a large amount of PV may cause excess energy during the day.

So, there are issues about the magnitude of PV that can be profitably used in a particular grid. Variability of sunlight may cause grid stability issues at large penetration of PV. These issues are not well understood at this time, and there is a need for hourly grid stability analysis using real weather from distributed sites over a region and hourly grid loads to understand how much PV can be used well. At this time, 10% PV in a grid seems to be an acceptable level without raising any grid performance issues [63].

Certain advanced PV technologies, particularly thin-film PV, use a variety of rare materials. These materials include indium, selenium, tellurium, and others. Resource constraints on these types of materials could restrict the ultimate deployment of some PV technologies. However, without considering new reserves, the known conventional reserves of these materials are sufficient to supply at least several tens of peak terawatts of thin-film PV [64]. For many other PV types, such as c-Si materials, supplies are virtually unlimited compared to any foreseeable demand.

Normally, rooftop PV is usually considered to displace the electricity used by the building where the system is located. There is usually more roof area available to produce electricity in a residence than what is used by the residence. To generate 100% of the electricity used in a typical home, the PV system should be somewhere between 2.5 and 6.4 kW. The roof area used by the PV system is between 230 and 640 ft^2 for the most popular single crystalline systems. The nationally average residential rooftop is about 1500 ft^2. The extra roof can generate more electricity than needed by the home and can be used elsewhere in the local grid. It is also possible to generate more daytime electricity than is needed by the local grid. Also, as PV achieves lower costs over the next decades, larger utility systems would become an option. The utility PV systems in more favorable solar areas can lead to exporting the excess electricity via transmission lines to more distant venues with potentially poorer solar resources [65]. For simplicity in the near-term time frame, we restricted this analysis to exclude these potential benefits of transmission and require all electricity generated from PV to be used locally.

Limiting PV during this time frame to 10% of the electric energy in a region to avoid questions of storage, cost is driven exclusively by geographic variability of resource. To generate a supply curve, the country is divided into 216 supply regions based on locations of recorded solar data [66] and calculated the cost of PV-generated electricity in each region, based on simulated PV system performance. The price is calculated in each region using PV system cost targets for 2005, 2015, and 2030. The amount of PV in each cost "step" is based on the capacity required to meet 10% of the region's total electricity demand. Each region's total electricity demand is estimated from population and state level per-capita electricity consumption data. Electricity consumption for 2015 and 2030 was based on projected national demand growth (Energy Information Administration, U.S. DOE, 2006) minus the energy efficiency potential under an aggressive carbon reduction policy [68].

Figure 4.26 below provides the resulting PV energy supply curve. The 2005 curve shows little usage due to the very high price of PV electricity (>25 cents/kW·h) even in relatively high-quality resource areas without existing subsidies. Because we only allow PV to provide 10% of a region's

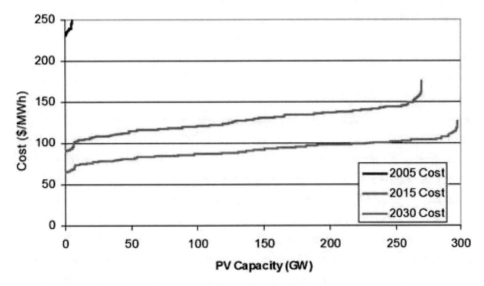

FIGURE 4.26: Roof top PV capacity supply. (from Ref. [68])

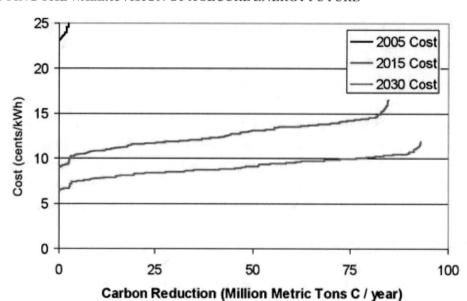

FIGURE 4.27: Carbon reduction curves for rooftop PV (based on 300 GW in 2030). (from Ref. [68])

energy, the absolute amount of energy is restricted in the 2015 and 2030 curves (note: 100 $/MW·h = 10 cents/kW·h). In reality, it should be possible to extend the supply curve to the right by considering scenarios involving transmission or considering possible cost impacts of intermittency. Including transmission could also potentially "flatten" the supply–cost curve, given the possibility of increased use of high-quality resources.

These restricted results indicate that about 300 GW of power, which is compatible with the growth and cost assumptions above. This amount of rooftop PV would generate about 450 TW·h/year of energy by 2030. Figure 4.27 translates these PV capacity values into potential carbon reduction.

It should be noted that the United States Photovoltaics Industry Roadmap goal is to deploy 200 GW of PV by 2030, and this is in the range of 150 to 300 GW used in the above analysis. This further limitation of PV is used to reflect the realities of the industry to assemble these systems. If it is assumed that the 200 GW of PV is uniformly distributed (based on population) across the United States, then this level of PV deployment would produce about 300 TW·h annually. The carbon emission reduction per year would be two thirds and in the range of 50–80 MMtC/year [68].

The basic resource potential for solar PV in the United States is virtually unlimited compared to any foreseeable demand for energy. Practically, deployment of solar PV on the scale necessary to have a significant impact on carbon emissions is contingent on a number of factors. PV is currently among the more costly renewable technologies but has significant potential for reduced costs and deployment on a wide scale. Sustained cost reductions will require continued R&D efforts in both the private and public sectors. Significant growth will be needed in the scale of PV manufacturing both at the aggregate and individual plant levels. For example, production economies of scale will likely require the annual output of individual PV plants to increase from a current level of tens or at most a few hundreds of MW per year to GW per year. Finally, institutional barriers, including the lack of national interconnection standards for distributed energy and net-metering provisions, will need to be addressed.

In the scenario we evaluated here, in which 200 to 300 GW of solar PV provides about 7% to 10% of the nation's electricity, PV is unlikely to be burdened by constraints of intermittency, transmission requirements, land use, or materials supplies. Available roof space (and zero to low-cost land) and known materials can provide resources far beyond this level of PV use. In the long term, at some point beyond this level of penetration, intermittency poses a more significant challenge, likely requiring large-scale deployment of a viable storage technology. However, as part of a diverse mix of renewable energy technologies, solar PV by itself can play a valuable role in reducing carbon emissions from the electric power sector [68]. These results will be combined with renewable energy options in Section 4.2.8.

4.2.4 Geothermal

There are currently 2800 MW of geothermal electric capacity in the United States, which provides about 0.5% of U.S. electricity. All of these plants and all geothermal power plants in the world use hydrothermal resources, which are naturally occurring reservoirs of hot water and steam located within a few thousand feet (or about a kilometer) of the surface. Most of the U.S. plants are located in California and Nevada.

Exploitation of future geothermal resources is focused on drilling to greater depths than today's plants. Figure 4.28 shows a map of temperatures at a 6-km depth, and it shows a geographically diverse energy source with most of the resources located in the west. A total of 16 states have temperatures exceeding 200°C, primarily in the western half of the United States.

As shown in Table 4.4, the energy content stored at 3- to 10-km depths in U.S. geothermal resources is vastly greater than the national annual energy demand [69]. In 2003, the total U.S. energy demand was 98 quads (1 quad = 1 quadrillion British thermal units), of which 84 quads were from fossil fuel sources (see EIA Home Page at http://www.eia.doe.gov/). The geothermal resource storage is about 14 million quads, which is about 140,000 times the total U.S. primary energy use. The scale of stored geothermal energy is so much larger than current demand that even very low geothermal energy recovery could displace a substantial fraction of today's fossil fuel demand. The sum of stored energy in-place plus steady-state conduction upward from the Earth's core could sustain foreseeable geothermal energy withdrawals over geologically long periods.

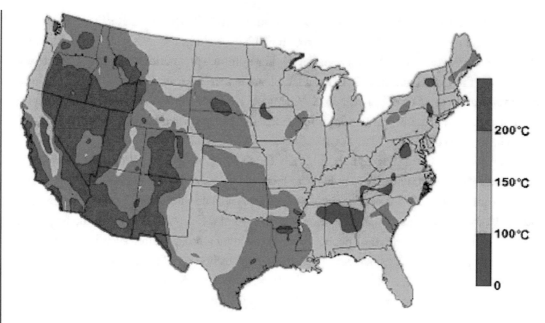

FIGURE 4.28: Geothermal temperatures at 6 km depth.

The challenge is using this vast baseload (24 h a day and 7 days a week) resource in an economical manner. The existing hydrothermal plants are all relatively close to the surface (within 1 km), and they all use hot water or steam from below the surface. This hot fluid is used to drive either a Rankine steam cycle or, for lower-temperature resources, a Rankine power cycle using a fluid with a lower boiling point than water, such as isobutane or pentane. The latter is called a "binary cycle" because there are two separate fluid loops. The first loop uses the heated fluid from the geothermal field, which transfers heat to the second fluid that is used in the heat engine-generator to produce power.

To reach and extraction heat from more remote resources within the Earth's crust in a competitive, commercial-scale manner is the challenge. It is not a technical limitation to reach depths of interest because of the availability of conventional drilling methods. The current technical issues are the economic uncertainty of site-specific reservoir properties such as permeabilities, porosities, in situ stresses, etc. For a geothermal field to be economically productive, it is necessary to set up a cluster of injection and production wells to access the reservoirs. Given the large potential of geothermal, the proportional payback for R&D gains is huge [70].

In the nearer term, the WGA Clean and Diversified Energy Study project has estimates that there will be about 6 GW of new power available from shallow hydrothermal resources by 2015 and a total of 13 GW available by 2025 (Clean and Diversified Energy Initiative Geothermal Task Force Report, Geothermal Task Force, Western Governors' Association, January 2006. Internet Report).

TABLE 4.4: Estimates of geothermal resource base, hundreds of quads (from Ref. [70])

RESOURCE TYPE	HEAT IN PLACE AT 3–10 KM
Hydrothermal (vapor and liquid)	20–100
Geopressured (including hydraulic and methane energy)	710–1700
Conduction Dominated EGS	
• Sedimentary EGS	> 1000
• Basement EGS	139,000
• Supercritical Volcanic EGS	740
Total	142,000

EGS, Enhanced Geothermal System.

By looking beyond shallow hydrothermal resources, the power potential increases significantly and these systems are called "enhanced geothermal systems." The enhancement can be in one of two forms: the expansion of existing hydrothermal reservoirs (sedimentary EGS) and accessing deep, hot dry rock (basement EGS). To access basement EGS, water injection under pressure is used to add both water and increase the permeability of the rock. A key opportunity is to use hot water from depleted oil and gas wells near the Gulf Coast [70].

To go beyond the Western Governors' nearer-term estimates for western hydrothermal geothermal power, Vorum and Tester estimated that a total of 100 GW would be available from the various resources by 2050 at costs of less than 10 cents/kW·h from these sources:

- 27 GW from hydrothermal
- 25 GW from sedimentary EGS
- 44 GW from oil and gas fields
- 4 GW from basement EGS
- 100 GW Total by 2050

Based on the use of binary cycles, costs for the two types of enhanced production are shown below expressed as LCOE in cents/kW·h.

TABLE 4.5: Estimated costs of geothermal power production		
REFERENCE CASE BASES	HYDROTHERMAL BINARY	EGS BINARY
Reservoir temperature (°C)	150	200
Well depths (km)	1.5	4
LCOE in ¢/kW·h		
LCOE as of 2005	8.5	29.0
LCOE as of 2010	4.9	
LCOE as of 2040		5.5

Table 4.5 shows the RD&D goals for enhanced hydrothermal systems using binary conversion at 150°C, approaching a DOE Program goal of 5 cents/kW·h in a near term at 2010. Also shown is the longer-term time frame for EGS binary systems around 2040.

Binary systems in use now generally have well depths under 1 km, and they accommodate temperatures marginally below about 200°C. The cost and magnitude estimates shown above are based on a well-organized DOE research program that would focus on technology opportunities in exploration, reservoir stimulation, drilling, and energy conversion to stimulate industry to transition toward a larger pool of resources. The RD&D goal is to be able to use a binary conversion system at temperatures down to 125–150°C in conjunction with well depths to 4 km (13,000 ft). The goal of functioning at the combined temperature and depth values is to expand the resource base for power generation.

Risk involved in geothermal power growth is due to both technical factors in finding a cost-effective resource to develop and a management barrier to commitment of funding at a predictable and competitive return on investment. Some of the issues here are up-front funding for resource exploration, engineering development, market (buyer) commitment, and commitment before-the-fact to installing transmission capacity for new power plants to access their prospective markets [70].

This geothermal energy introduction versus time is evaluated versus other conventional energy sources in a national energy market using the National Energy Modeling System. The resulting geothermal energy supply curves are shown in Figure 4.29 with the cost expressed as the LCOE in cents/kW·h. About half of the projected 100 GW would be on-line in 2030 assuming that the DOE R&D program to reduce costs is carried out. Without the DOE program, the geothermal power in 2030 would drop to about 35 GW.

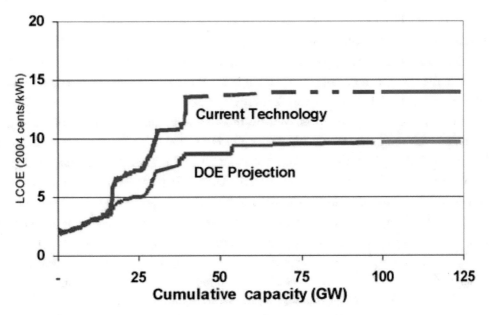

FIGURE 4.29: Geothermal supply curves. (from Ref. [70])

Assuming climate change concerns spur continued DOE research to lower costs and the 50 GW level is achieved by 2030, the carbon displacement is 63–103 MtC/year. The midrange value is 83 MtC/year. It should be noted that in the absence of a DOE program to reduce costs, this would drop to 30–35 GW [70].

As in the case of biomass electricity, a geothermal plant runs 24 h per day, seven days per week, and can provide baseload power, thus competing against coal plants. Thus, the high-end value that assumes substitution for existing coal plants may be realistic for geothermal, although the midrange value is used in the summation.

Geothermal energy is an existing baseload power system using well-established commercial systems. Although it currently provides about 0.5% of the total U.S. electricity, it has enormous future potential to generate significant carbon-free power. A relatively well-distributed resource with a Western emphasis, the future reality depends on extending existing geothermal fields and accessing deeper reservoirs. The success of these efforts depend on a continued R&D program to remove much of the risk involved in extending existing oil/gas field technology into geothermal applications. A number of independent estimates [69] conclude that about 50 GW of baseload geothermal will be on-line at a competitive energy price (<5 cents/kW·h) by 2030. This amount of carbon-free energy translates into replacing 70 GW of coal-fired power. The expected carbon emission reduction is about 63 to 103 MtC/year. These results will be combined with renewable energy options in Section 4.2.8.

4.2.5   Biomass Power

Biomass is part of a closed carbon cycle, and it can reduce CO_2 generation when used as a substitute for a fossil fuel. That is, the CO_2 generated when biomass is combusted is absorbed from the atmosphere when the next generation of biomass is cultivated if the crops are grown sustainably. Biomass currently provides about 10% of the world's energy, and two thirds of this is called traditional biomass and used in developing countries. Modern biomass use in industrialized countries is usually for combined heat and power for electric power generation and as a transportation fuel. About 40 GW of electric capacity is in place globally, and about 10 GW is in the United States and growing at 3% to 4% per year. The major current use in the United States is for industrial process heat as well as biomass-based transportation fuels use. The biofuel sector is growing at 18% per year due to high oil prices and large ethanol supports. The key feature of the current bio-liquids program is that it uses a food, corn kernels, as the feedstock to generate ethanol. However, biomass can be collected as a residue from current biomass commercial systems or grown as an energy crop specifically for displacing carbon. In addition, biomass can be used to sequester carbon in the plants themselves if done on a long-term basis (see Section 4.2.5.3, Biomass Sequestration).

The amount of carbon offset achieved by biomass use depends on the particular way that the biomass is used to provide the energy service and the fossil system that the biomass is replacing. A recent study calculated that there can be an enormous difference in carbon displacement even when you are starting with the same source of biomass. This difference can be as large as a factor of 12 between the most and least effective way to use biomass as shown below [71]. When results are measured in tons of carbon offset per hectare of land per year (tC/ha per year), the most effective system is to use a lignocellulosic crop in an efficient electric generation plant such as one using an IGCC where this biomass system displaces a traditional pulverized coal power plant. When this biomass is used in this way, 6.3 tC/ha per year are displaced. A high-yielding lignocellulosic crop would be short-rotation woody crops such as willow or poplar and herbaceous energy crops such as switch grass or miscanthus. This very same crop used in the same way is only half as effective and displaces 3 tC/ha per year when compared to a natural gas power plant instead of coal. This is so because the natural gas plant uses carbon about twice as efficiently to generate electricity than a pulverizing coal steam boiler power plant.

If the lignocellulosic crop is converted to ethanol and used to displace petroleum in an internal combustion engine (ICE), it displaces 1.1 tC/ha per year, which is nearly six times less effective than generating electricity efficiently and displacing coal. Also, the lignocellulosic is about 38% more effective than using a cereal crop such as wheat to generate the ethanol (1.1 compared to 0.8 tC/ha per year). The lignocellulosic crop can be grown on more marginal land without fertilizers and is not a food crop. If the lignocellulosic crop is converted to hydrogen and compressed and used in a vehicle, it displaces 4.6 tC/ha per year compared to using petroleum in an ICE. This is four times more effective compared to converting the lignocellulosic crop to ethanol and is a powerful impetuous to developing a commercial and cost effective hydrogen vehicle power system.

The bottom line is that if you want to reduce carbon emission using a nonfood biomass as a feedstock, then the most effective way to do it is to burn the biomass in an efficient IGCC electric power plant to offset using pulverized coal to generate electricity (6.3 tC/ha per year). This electricity can be used to power a pluggable hybrid-electric vehicle and then displace the use of petroleum. The next most effective way to use a nonfuel biomass is in a fluid bed central station to generate and compress hydrogen for use in fuel cell-powered vehicle to displace the use of petroleum in an ICE (4.6 tC/ha per year) (Table 4.6). This is a longer-term prospect because is requires that the infrastructure to distribute hydrogen is in place and that cost-effective fuel-cell vehicles are

BIOMASS SOURCE	**ENERGY SYSTEM**	**FINAL ENERGY FORM**	**FOSSIL OFFSET**	**CARBON OFFSET (TC/ HA PER YEAR)**
Lignocellulosic biomass	IGCC	Electricit	Pulverized coal boiler	6.3
Lignocellulosic biomass	Central plant compressed H2	H2 in fuel-cell vehicle	Petroleum ICE	4.6
Lignocellulosic biomass	Fluidized bed combustor steam power plant	Electricity	Pulverized coal boiler	3.8
Lignocellulosic biomass	IGCC	Electricity	Natural gas turbine	2.9
Lignocellulosic biomass	Central plant	Ethanol	Petroleum ICE	1.1
Cereal crop (wheat)	Central plant	Ethanol	Petroleum	0.8
Oilseed (soya, rape)	Central plant	FAME	Petroleum ICE	0.5

TABLE 4.6: Carbon displacement with different crops and energy systems

FAME, fatty acid methyl esters.

available. The current approach using a food source of biomass to generate ethanol to displace petroleum in an ICE is nearly 10 times less effective at reducing carbon emissions.

4.2.5.1 Biomass Resource. At present, biofuels are produced from sugar cane (Brazil) and corn (United States). Recent attention in the United States is directed at gathering biomass from residues from urban centers, agricultural residues, forest residues, mill and process residues, and residues from food processing and animal husbandry. A recent compilation of these biomass residues is shown in Figure 4.30. This atlas shows county-level resource technical potentials for all the biomass sources identified above [72]. The greatest intensity of these residue biomass resources is in the Corn Belt and in urban centers. The urban center residual biomass is where the people are and is distributed primarily in the center-eastern half of the county and on the west coast [73].

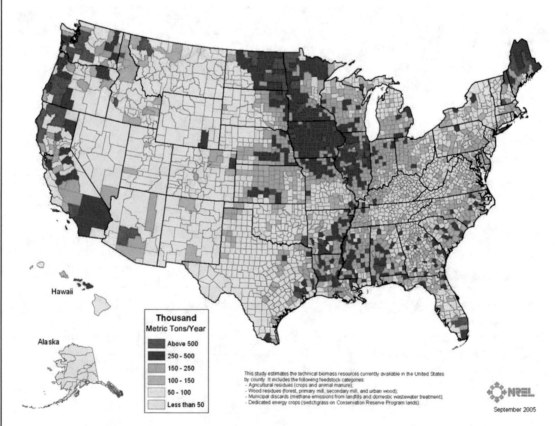

FIGURE 4.30: Biomass residue resource distribution by county.

Two key studies are available that provide an initial basis for estimating economically accessible biomass. The "Billion Ton Study" is a collaborative effort among the U.S. DOE, the U.S. Department of Agriculture, and the national laboratories led by the Oak Ridge National Laboratory [74]. This study estimates a potential 2025 crop and biomass residue contribution of 1.26 billion metric tons (all biomass is described on a dry basis).

The second major study is the WGA Clean and Diversified Energy Study by the Biomass Task Force [75]. This study generated an electrical energy supply curve for the WGA 18-state region that identified slightly less than 50% of the region's technical potential as being economically accessible to generate 15-GW electricity at less than 8 cents/kW·h. These two studies can be combined to extend the more detailed WGA study to the entire county [73].

4.2.5.2 Biomass Electricity. To understanding the magnitude the role that the use of biomass residues in generating electricity could play in displacing carbon, it is necessary to generate a biomass resource supply curve. The different biomass feedstocks need to be collected, brought to a central location, and used as a heat source for the appropriate power plant. To do this, various studies need to be combined such as in the SHAM model from Nilsson where supply curves are generated for the delivery of a raw biomass feedstock to a central location for processing [76]. The two most important factors in establishing a biomass supply curve versus cost are the annual yield of biomass per area of land and the efficiency of the process of converting the biomass to the final energy product.

Traditionally, a supply curve—a plot of cost per unit of output (cents/kW·h of electricity) against the cumulative output—would be generated for a specific plant location, and the size of the plant would be determined by the output below some cost threshold. Because the plant location and the potential biomass resource generation locations would be fixed, the actual road network would be used for the logistics. Likewise, the farm or forest production costs would be based on best practice in the region, taking into account the local climate and soil determinants of plant productivity. At the conversion process plant, the investment cost and those of the fixed and variable operating costs would be determined from vendor quotes.

A working group determined a supply curve for the 18 western states for the WGA Clean and Diversified Energy Initiative (CDEAC) in which a series of approximations were used to derive a supply curve for the region. The working group used the best available data sources on quantity, quality, environmental constraints, and cost of agricultural, forest, and urban biomass resources. At this stage in this analysis, dedicated crops could be considered to be equivalent to a high yield of agricultural residue; and their planting would, in effect, displace agricultural crop plantings generating residues in specific areas. The major data sources were compiled to include the following biomass resources: agricultural residue (lignocellulosic biomass that remains in the field after the harvest of agricultural crops), forest-derived biomass, urban residues usually called municipal solid waste, and landfill waste in place where landfill gas is converted to heat or electricity [71].

A logistic model was developed by Overend and Milbrandt to deal with the difficulties of conducting a search of each county to locate the resources, power lines, and substations to optimize the transportation costs of moving the harvested and collected biomass to the conversion plant and transmitting the electricity to the grid. To facilitate the analysis, they used a simple logistics model that assumed that the resource area was contiguous, and they estimated the fraction of the land area devoted to the resource from the area planted with the crop(s) chosen (wheat and corn) relative to the total land area [77].

The county-level productivity per unit area of the crop was then used to compute the total metric ton per kilometer transportation effort required. For agricultural land, the assumption made was that a plant would be centrally located in the arable land area. For forest crops, the conversion plant was assumed to be at the center of the diameter chord for a semicircular collection area. Based on the *Statistical Abstract of the United States (1997)* [78], a national average local freight charge per ton-mile of approximately $0.24 was estimated. Based on this and the Billion Ton Study, a base case of $0.20 per green ton-mile was used [73].

The biomass is converted into an intermediate product (heat or fuel gas) that is then converted into electricity. Although there are a large number of potential process configurations, the following three processes are used to simplify the analysis to generate the desired supply curve:

- Stoker and fluid bed combustors with steam generation and steam turbines
- Gasification with applications to boiler steam generation and steam turbines, or abiomass combined cycle (gas turbine, heat recovery steam generator, and steam turbine), or an ICE
- Anaerobic digestion (animal, water treatment, and landfill) with ICE or gas turbine.

The capital cost (CAPEX - capital expenditure) of these plants is shown in Figure 4.31 as a function of plant size in MW [75]. For each supply resource, the levelized cost of electricity was calculated depending on the cost of harvest/collection, transportation, and conversion technology using the NREL methodology [79]. Discount factors and inflation factors supplied by the WGA as described in the CDEAC guidance documents [80].

The maximum size of a biomass facility (either stoker combustor or IGCC) would be 120 MW and would be connected to the high-voltage distribution grid via substations. Units that were below 60 MW would be connected to the local distribution grid at local substations. For counties with biomass supplies greater than the largest plant size, multiple plants were assumed to use up the available supply. For any given resource supply and cost, the generation technology selected was

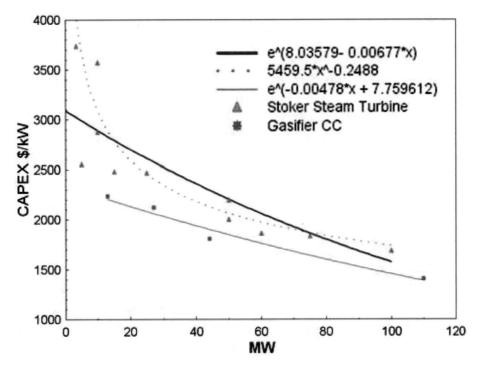

FIGURE 4.31: Capital cost (2005 dollars) versus size of power plant.

the one that resulted in the lowest-cost outcome. For larger sizes, this tended to be the IGCC; for the smaller units (less than 15 MWe), the technology would either be a stoker steam turbine or a gasifier–ICE combination. The resulting supply curve was summed over seven biomass sources as shown below for the 18 western states (Figure 4.32).

The results are that 15 GWe of baseload generating capacity would be available at less than 8 cents/kW·h (80 $/MW·h). This 15 GWe is based on the 170 Mt of biomass in the WGA study. This result is then projected to the 1.26 Gt of biomass in the Billion Ton Study for the nation. Extrapolation from the 170 Mt of biomass in the WGA study to the 1.26 Gt of biomass in the Billion Ton Study produces an estimate of 110 GW of electricity and a carbon offset of between 139 and 225 MtC/year.

The 15 GWe from the Western states would annually reduce carbon emission by 31 million metric tons of carbon using a conversion factor of 260 tC/GW·h. Extrapolating this result to the entire county results in an average carbon emission reduction of 183 MtC/year. This is achieved at an estimated electricity-generating cost range of 5 to 8 cents/kW·h [73]. This result is combined with all the other renewable energy options and the energy efficiency options in Section 4.2.8.

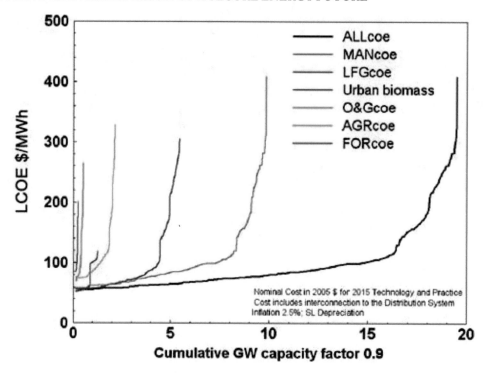

FIGURE 4.32: Biomass electric generation supply curve for western states. All, the sum of all six sources; Man, Manure; LFG, Landfill Gas; Urban Biomass, Municipal Solid Waste; O&G, Orchard and Grapes (California only); AGR, Agricultural Residues; FOR, Forestry Resources. (from Ref. [73])

4.2.5.3 Carbon Sequestration. A form of biomass not included in this study is the sequestration of carbon in sustainable biomass. One form of this is to increase the residence times of carbon in soil. Quantification of the transformation pathways and residence times are subjects of intense study to clarify the role of soil carbon in climate change mitigation [81–83]. Changes in agricultural practice such as no-till agriculture, which eliminates soil disturbance by plowing and cultivation, also impact the ability of soils to fix carbon under different cropping regimes. Typical rates of carbon fixation are on the order of 0.2–0.5 tC/ha per year compared with offset rates from the table above for power systems on the order of 2.9–6.3 tC/ha per year for biomass to electricity. Because these techniques for sequestering carbon are about 13 times less effective per hectare, they are not considered in the study. However, we have enormous amounts of land under crop cultivation (~440 M acres), and the change of agricultural techniques might have a large impact of carbon even if the improvement per acres is smaller than those of focused new techniques.

Another form of carbon sequestration in biomass is to maintain forests and other flora to avoid the release of carbon when these current uses are converted to land uses that bind less carbon, or to replace some current land uses with forests or other flora that increase the carbon bound to the new biomass on a sustainable basis. "Organic farming offers real advantages for such crops as corn and soybeans compared to convention farming methods. The fact that organic agriculture systems also absorb and retain significant amounts of carbon in the soil has implications for global warming," says David Pimentel, professor of ecology and agriculture at Cornell University, "pointing out that soil carbon in the organic systems increased by 15–28%, the equivalent of taking about 1000 lb of carbon per acre out of the air" [84]. "Organic farming approaches for these crops not only use an average of 30% less fossil energy but also conserve more water in the soil, induce less erosion, maintain soil quality and conserve more biological resources than conventional farming does," Pimentel added.

The Pimentel study compared a conventional farm that used recommended fertilizer and pesticide applications with an organic animal-based farm (where manure was applied) and an organic legume-based farm (that used a 3-year rotation of hairy vetch/corn and rye/soybeans and wheat). The two organic systems received no chemical fertilizers or pesticides. Organic farming can compete effectively in growing corn, soybeans, wheat, barley, and other grains, but it might not be as favorable for growing such crops as grapes, apples, cherries, and potatoes, which have greater pest problems.

Carbon sequestration in agricultural soils in the United States has the potential to remove considerable amounts of carbon from the atmosphere annually, with estimates ranging from 75 to 208 million metric tons annually and representing nearly 8% of total emissions of the United States [85].

Studies show that soil sequestration is competitive with other strategies for carbon mitigation, such as reforestation and carbon offsets associated with use of biofuels, but the quantity of carbon sequestered will depend on the price for carbon credits [86, 87].

The sequestration potential of conservation tillage is smaller than that of perennial grasses, but the crops that were being grown before with conventional methods can continue to be grown with greater carbon sequestration using organic techniques [88].

The framework described above is applied at the county level for the state of Illinois. The crop choices included are four row crops (corn, soybeans, wheat, sorghum) grown using either conventional or conservation-tillage practice and three perennial grasses, pasture, a forage crop, and switchgrass and miscanthus, as two bio-energy crops that can be co-fired with coal to generate electricity at existing electricity-generating plants in Illinois.

The result of the absence of any carbon targets or bio-energy subsidy is that 45% of the 23.2 million acres of cropland in Illinois would be under conservation till. As a result of conservation tillage and pasture, the soil carbon level increases by 23 MtC/year by 2017; 93% of this increase is due to use of conservation tillage.

If these results were applied to the entire country, the amount of carbon sequestration in the soil due to conservation tillage would be 438 MtC/year. This is substantially more than the estimated 183 MtC/year achieved by diverting biomass wastes to electricity production by 2030. Thus, there is enormous potential to converting conventional agriculture technique to organic conservation tillage. The only issue is that for this to be counted toward carbon emission reduction, this transition to conservation tillage must be sustained for the long term. It is like carbon sequestration in underground aquifers, if there is even a leak rate of 0.1%/year, the concept goes down the drain. Similarly, if 0.1% per year of the land devoted to these sequestration techniques is diverted to other uses, the long-term usefulness of these techniques is doubtful.

How we depend on the permanence of conservation tillage in U.S. agriculture or the status of a reforestation program is a difficult issue. We would have to maintain a policy that supports this transition for the long term (centuries). A reversal to techniques that do not stabilize as much carbon would erase all the merits of this option. The conservation tillage option along with reforestation should be pursued, but its potential is not considered in this study.

4.2.6   Biofuels

If your goal is to reduce carbon emissions, it is nearly 10 times more effective to use biomass to generate electricity than to use it for transportation liquids derived from corn (see previous section). However, there are powerful reasons to use some biomass to create transportation liquids, and this is considered as part of an overall strategy to meet both our need to reduce carbon emission and begin to deal with oil importation problems.

Transportation uses 150 Bg/year (billion gallons per year) of gasoline and this contributes about 32% of the U.S. carbon emissions [89], and this sector's contribution to carbon emissions has been steadily growing over the last decades. Although the combustion of a biomass-derived liquid fuel will generate CO_2 when combusted in an engine, this CO_2 is recycled between the atmosphere and the biosphere as new crops are grown with a net CO_2 contribution of zero.

There is an active biofuels industry in the United States at this time and 6.5 Bg of fuel ethanol were generated in 2007 [90]. This resulted primarily from the fermentation of starch in corn grain to ethanol. This industry has been growing at 18% per year over the last decade. Recent enacted Renewable Fuel Standard establishes a mandate to reach 7.5 Bg/year by 2012. It appears that this level of production will be reached in 2008.

A newcomer biofuel is biodiesel from a vegetable oil-based diesel fuel substitute. This uses a relatively simple thermochemical process where the natural oil is chemically combined with methanol to form FAME (fatty acid methy esters). Biodiesel production tripled in 2005 to reach 75 million gallons, which is about 2% of the ethanol production.

Although important as an initial step toward using biomass for liquid transportation fuels, ethanol from corn is limited by two factors. The first is the heavy use of petroleum products to grow corn. For each gallon of ethanol produced from corn, the change in carbon emissions compared to an equivalent amount of gasoline is a reduction of only 18% on average. The range of values in the literature goes from a high of producing 21% more carbon than gasoline to a low generation of 53% less than gasoline [91]. Although a biofuel could potentially be carbon-neutral, when all the fossil energy is accounted for in the growing of the corn, gathering, and the ethanol-generating process, there is a carbon savings of only 18% on average.

The second factor limiting corn is that corn is a food and its use as a transportation fuel feedstock would eventually unduly impact the price of food. A limit must and should be established for corn-based ethanol and it may be less than the mandated 7.5 Bg/year. It should be less than the amount of corn that starts to impact food prices. There is some indication that the current production of corn to ethanol is already impacting the cost of corn and partially responsible for increased world hunger. A recent formal study by Keith Collins of the Department of Agriculture suggests the increased corn demand for ethanol could account for 25% to 50% of the corn price increase with ramifications throughout the food production sector [92]. Thus, it appears that at the 2007 production level of 6.5 Bg/year, we may have already exceeded the reasonable limit for corn-to-ethanol production. A gallon of ethanol contains about two thirds of the energy of a gallon of gasoline. 6.5 Bg/year is about 4.3 Bg/year of gasoline equivalent and is about 3% of the current U.S. demand of 150 Bg/year.

Ethanol made from agricultural residues and switchgrass is about 4 times more effective than corn-based ethanol. The GHG emission reductions from switchgrass is about 2050 grams of equivalent per gallon of gasoline equivalent, while corn ethanol saves only 550 grams [93, 94]. Thus, developing commercial processes that use switchgrass as the bio-feedstock is many times more effective at reducing carbon emissions and eliminates the competition with foods. This area should be the primary focus of RD&D efforts in the biofuels area.

To avoid some of the ambiguities and complexities of the entire biomass supply, this study focuses on two of the largest and most straightforward elements of the supply—crop residues and perennial grasses grown for energy production. Neither of these feedstocks is produced or used in large amounts today. Their introduction into the supply will most likely be driven by the demand for energy from biomass and not demand for food. This source of biomass for liquids is almost mutually exclusive from the biomass used in the previous chapter for electricity generation [95].

There are several processes that have been studied for decades to convert lignocellulosic biomass into fuels. These include thermochemical processes such as the use of the well-established

Fischer–Tropsch process that uses gasified biomass to generate liquids via a catalytic conversion; or pyrolysis of biomass to produce bio-oils that can be used in petroleum refining. The other type of approach is biological processes such as enzymatic hydrolysis of cellulose combined with fermentation of sugars and the use of residue for heat and power production [95].

To simplify analysis, only one conversion technology is considered—the biological conversion of lignocellulosic biomass to ethanol and excess electricity. A key to this process is the use of biological catalysts that can break down the carbohydrate polymers in biomass to sugars and microbes that can ferment all the sugars in biomass to ethanol in a cost-effect manner. The current DOE Biomass Program goal is to achieve a nominal minimum selling price of ethanol from biomass of $1.07/gal of ethanol by 2012 [96]. This goal moves to $0.62/gal by 2030 by combining enzyme production, hydrolysis, and fermentation capability in one microbe [97]. The technology target cost for biological conversion of lignocellulosic biomass to ethanol (gasoline equivalent) and the range of EIA predicted gasoline cost are shown in Figure 4.33 [95].

Assuming the DOE RD&D program has some success, the wholesale cost of a gallon of ethanol in 2015 could be $1.07/gal and also assuming the cost of $33 per dry ton of biomass delivered to the facility and an average yield of 99 gal/t of biomass. When corrected for energy equivalency to gasoline, it becomes $1.62/gal wholesale price ($68/barrel) as shown in Figure 4.33. This can be compared to the 2015 EIA wholesale gasoline price estimate of $63/barrel (middle range) to $96/barrel (high range) [95].

It is likely that ethanol would become cost competitive with oil well before 2015 since the EIA projections are unduly conservative. One can observe that the price of oil in early 2008 is already well over $100/barrel. Also, over the last two decades, the price of oil has risen by an average rate of 7%/year in current dollars. The EIA prediction of oil prices raising only 2%/year would seem seen to support the claim that EIA oil price predictions are overly conservative. This especially true when global oil production peaking this decade is considered (see Chapter 2). The high range oil price estimate closer to $100/barrel is more likely and the cost of lignocellulosic biomass/ethanol of $68/barrel would strongly encourage ethanol introduction into the market. However, the EIA low ball estimates are used to introduce a large degree of conservatism into these projections and this would result in the bio-liquids program cost goals and the EIA gasoline price projections become equal around 2015.

If the government sponsored corn/ethanol program is continued, it should level out at a produce level of about 6.6 Bg/year of subsidized ethanol. The resulting cost of the biofuels program using residue and switchgrass would be close to the midrange EIA oil price predictions and no additional production of ethanol would result. Using the high-range of EIA oil price predictions, a total of 14 Bg/year by 2015 as shown in Figure 4.34 [95]. This is more than 9% of current U.S. oil consumption when the ethanol is placed on a gasoline equivalent basis.

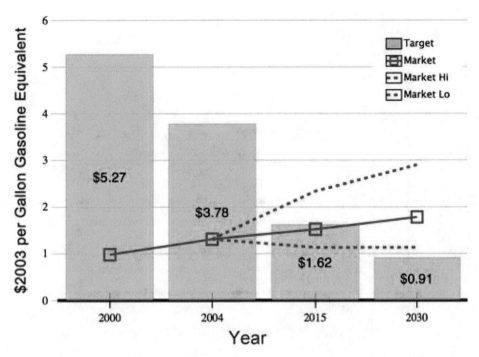

FIGURE 4.33: Technology target cost for biological conversion of lignocellulosic biomass to ethanol compared to future gasoline price. (from Ref. [95])

Repeating this analysis for 2030, the wholesale cost of ethanol would be $0.62/gal assuming the RD&D program success and a biomass feedstock cost of $44/ton and an average yield of 116 gal/t of biomass. This is equivalent to $0.93/gal of gasoline ($39/barrel). The EIA midrange estimate of the 2030 gasoline wholesale price would be $1.80/gal ($76/barrel), and $2.90/gal ($122/barrel) on the high range. Ethanol cost would clearly be economic at half to one-third the expected oil price in the 2030 time frame. The expected bio-liquid production would be about 28 Bg/year including the 6.6 Bg/year of subsidized corn as shown below [95]. This is about 18.7% of current oil use. If the 2030 price of oil is assumed to be more than $2.25/gal, then the total ethanol production would be 35 Bg/year, which is more than 23% of current gasoline use.

Biofuels made from agricultural feedstocks such as corn, agricultural residues, and energy crops could play an important role in reducing carbon emissions. Using the greenhouse gas reduction of around 2050 grams of equivalent per gallon of gasoline equivalent for lignocellulosic biomass/ethanol and corn/ethanol reduction of 550 grams translates into a carbon reduction of 4 MtC/year in 2015 and 58 MtC/year in 2030 as shown below [95]. The wholesale price as gasoline

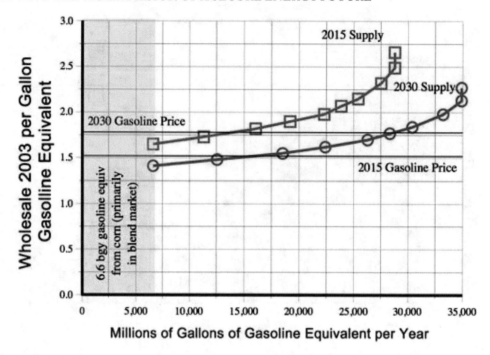

FIGURE 4.34: Ethanol supply curves in 2015 and 2030. (from Ref. [95])

shown in these figures is the midrange EIA estimate. This carbon reduction could be up to 70 MtC/ year if the projected oil price is greater than the EIA estimates.

With the projected increase in efficiency shown in Sections 5.1 and 5.4.3, about half of the expected 2030 total oil use (10 Mbbl/day) will be met by efficiency improvements. EIA predicts that most of the expected growth in liquid fuels will be in the transportation sector and estimates that use will go from 14.9 to 20 million barrels per day by 2030 (Figure 4.35). The expected energy efficiency improvements will keep the total transportation liquids at about today's level so that the biofuels contribution will be about 37% to 47% of the 2030 gasoline demand. The range depends on whether the expected price of oil is the EIA midrange or high range. This is considerable contribution to the U.S. gasoline demand and will significantly reduce the amount of imported oil as well as reduce the carbon emission from the vehicle transportation sector.

4.2.7 National Transmission System

The estimates of installed capacity of CSP, geothermal, and wind power plants with the resulting carbon reduction by the year 2030 assume severe transmission restrictions on both energy systems.

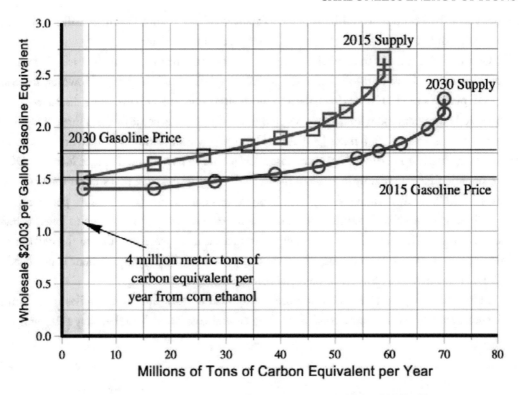

FIGURE 4.35: Carbon reduction from biofuels in 2015 and 2030. (from Ref. [95])

Limited available capacity on existing transmission lines are assumed, and when this transmission capacity is used up, then the renewable power plant is assumed to build new transmission lines to nearby load centers (cities) using the simplifying assumption of $1000/MW/mile. CSP, wind, and geothermal energy resources are significant national energy sources with potential power levels many times today's U.S. electric capacity. Limiting these technologies to regional energy supply by the transmission assumption is adequate up to the year 2030. It is more likely that after serving regional needs, wind, geothermal, and solar concentration power sources would use long-distance transmission to bring this regional energy source to load centers around the country. This means that to capture the potential of these renewables, the United States would make a transition from a relatively fragmented regional electric grid system to a national electric transmission system.

High-voltage DC (HVDC) transmission lines would be used to connect the regional Southwest CSP to major load centers in the rest of the country. Transmission distances from 800 to 2500 miles would carry energy to all parts of the county outside the regional loads of the Southwest. The wind and geothermal resources are more widely distributed so transmission distances to bring their electricity to urban centers would be shorter. Distances of less than 1000 miles will allow excess regional wind and geothermal to be moved to other parts of the country.

Conventional electric transmission lines are AC systems primarily for the ease of changing voltage as you move from transmission to the multilayered distribution system. However, there are many examples where HVDC has been used to move large amounts of power long distances. There are about 70 DC transmission links that exist in the world today and some of them move 3000 MW of power for a distance of 1000 miles [98]. An AC to DC inverter station is needed at the start and end of a long distance DC link. This extra cost is overcome as power and distance increase, DC is more economic than AC mainly because of the reduction in line cost due to a smaller tower (hence, less steel work); a smaller "right-of-way" requirement; less conductors (two bundles per circuit instead of three); and each wire bundle may comprise fewer parallel conductors due to lower voltage stress. In addition, AC transmission over longer distances is more expensive due to the need or the addition of either series or shunt reactive power compensation. Such a requirement is not needed for HVDC transmission (Carl Baker, personal communication).

The distance where DC becomes more economical is from 200 to 400 miles in transmission length depending on specific conditions. The capital cost of both high-voltage AC and DC

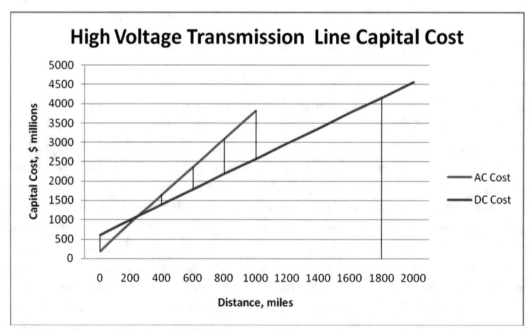

FIGURE 4.36: High-voltage transmission line capital cost.

is shown in Figure 4.36 below for transmission lines that carry 3000 MW [99]. It should be noted that there is variability in capital cost due to a number of factors such as terrain and environmental impact issues and mitigation, and that these figures are representative.

The energy cost for this transmission system including the line losses is shown in Figure 4.37 assuming that the cost of the generated electricity is 10 cents/kW·h. An AC transmission line at 300 miles adds about 1.25 cents/kW·h to deliver the electricity to the load center. AC transmission becomes increasingly expensive at greater distances and would add about 3.25 cents/kW·h to the delivered cost of electricity at 1000 miles.

DC transmission is similar in cost to AC at 300 miles but is less expensive at greater distances. At 1000 miles, DC would add about 2.25 cents/kW·h to the base electricity cost. Based on this representative analysis, using DC transmission to bring energy 1000 miles is only about 1.0 cent/kW·h more than using AC at 300 miles. At 1000 miles, DC transmission increases the cost of the delivered electricity about 20% and gives distant cities access to large and inexpensive renewable resources.

In addition to existing HVDC technology, superconducting electricity transmission holds the promise of reducing transmission losses by half at large distances. This promise must await the results of successful RD&D in this area.

4.2.8   Summary of Carbonless Energy Options
The major findings of this section are that:

- Energy efficiency could negate U.S. carbon emissions growth.
- Renewables can provide deep cuts in emissions.
- This combination can achieve a right on target carbon emission reduction of 1200 MtC/year by 2030 which is right on target to meet the U.S. goal of about 80% reduction by 2050.
- U.S. wind can provide approximately one-third of the renewable energy contribution and the remaining renewable contribution is split about evenly among the five renewable technologies.
- The United States has abundant renewable resources distributed throughout the country.

The bottom line is that the total contribution from these nine carbonless techniques for the year 2030 is about 1,200 MtC/year. This is on target to achieve carbon emission reductions of about 80% from today's value by 2050.

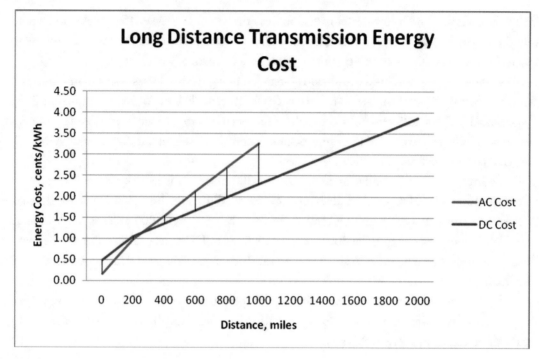

FIGURE 4.37: Long distance transmission energy cost.

4.2.8.1 Carbon Emission Reduction. The use of each renewable option was well below its potential and limited by a realistic concerns in the 2030 time frame. For example, wind is limited to 20% of grid electric generation in 2030 due to the assumption that there was limited access to existing transmission capability. And the PV potential was limited by the industry-estimated production capability of 200 MW in 2030. One area where we must avoid double-counting is with biomass and biofuels. Although converting biomass to electricity provides the greater carbon reduction, there is a strong national interest in directly displacing foreign oil. So for the sake of this analysis, we will assume that biomass for liquid fuels take precedence over biomass for electricity. The biofuels study was based on the use of some crop residues and energy crops and resulted in a 58 MtC/year carbon emission displacement. If we assume that the biomass dedicated to liquid fuels is removed from the types of biomass in the projected 1.25 billion metric tons used in the biomass study, we are left with 41% of that biomass available to produce electricity. Using all the biomass to produce electricity provided a carbon displacement of 183 MtC/year, and 41% of this yields 75 MtC/year. All the renewable technologies contribute a total reduction of 523 MtC/year of carbon emission from the atmosphere by 2030.

Energy efficiency improvements can be viewed either as lowering the BAU curve or as a wedge of displaced carbon. We will use the result of the overall energy efficiency study because this dealt with energy savings from efficiency improvements in electricity, natural gas, and oil using a reasonably consistent methodology. Translating these energy efficiency results into carbon emissions by using the average of the lower (national electric mix) and upper (coal) displacement, the carbon savings is 688 MtC/year by 2030. The total of all nine carbonless options is 1211 MtC/year as shown in Table 4.7 [100].

The purpose of this study was to consider a renewables-only scenario that focuses on what renewable energy can do in the absence of any new nuclear or coal gasification (with carbon capture) plants. These nonrenewable options are a potential means for addressing climate change, but they require longer lead times than the renewable options and they present other environmental and economic problems (see Chapter 5). The costs of new nuclear plants and coal gasification plants with carbon capture and storage will likely be sufficiently high that renewables will be very competitive economically.

These results are also shown in Figure 4.38 [100]. Approximately 57% of the carbon reduction contribution is from energy efficiency and about 43% is from renewables. Energy efficiency

TABLE 4.7: Carbon reductions from energy efficiency and renewable energy options	
CARBONLESS OPTION	**POTENTIAL CARBON REDUCTIONS (MTC/YEAR IN 2030)**
Energy Efficiency	688
CSP	63
PVs	63
Wind	181
Biofuels	58
Biomass	75
Geothermal	83
Total	1211

measures can allow U.S. carbon emissions to remain about level through 2030, whereas the renewable supply technologies can provide large carbon reductions.

The relative contributions of different renewable energy technologies are shown in Figure 4.39.

The total contribution from these nine carbonless techniques for the year 2030 is between 1000 and 1400 MtC/year (with a midrange value of about 1200 MtC/year). This is on target to achieve carbon emission reductions of between 60% and 80% from today's value by 2050. The carbon reductions in 2015 range from 375 to 525 MtC/year, with a midrange value of 450 MtC/year.

How much renewable electricity does this represent relative to what is needed? The current U.S. annual electric output is 4038 TW·h, and the EIA BAU projection is a value of 5341 TW·h by 2030, of which 4900 TW·h is from fossil fuels. The energy efficiency chapter estimates an an-

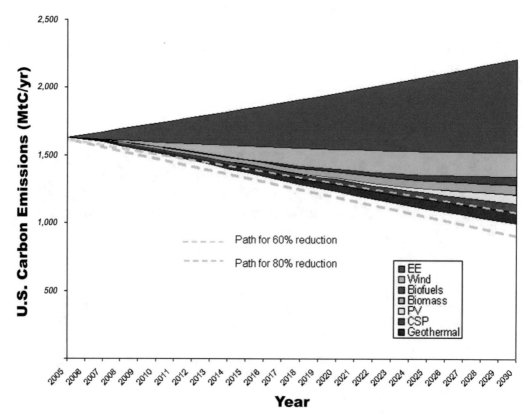

FIGURE 4.38: Path of potential carbon reductions in 2030 from energy efficiency and renewable technologies compared with reductions of 60% and 80% below today's emissions value by 2050 (from Ref. [100]).

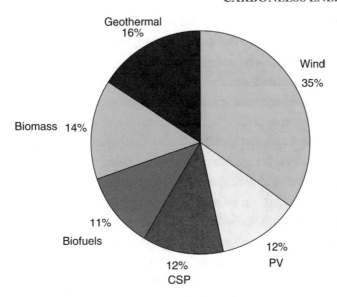

FIGURE 4.39: Relative contribution to carbon reduction in 2030 among renewables considered (from Ref. [100])

nual savings of 980 TW·h in 2025. Using the EIA BAU growth rate (1.2%/year) to extrapolate, we project an energy saving of 1038 TW·h in 2030. This leaves a total electric energy generation of 4303 TW·h (5341–1038) in 2030. Table 4.8 lists the annual electricity generation in TW·h for the various renewable energy technologies.

Summing the renewable electricity contributions results in about 50% total grid penetration in 2030. This is significantly higher than a commonly stated goal of "30% by 2030," but this may not account for a reduction in electric energy production from aggressive efficiency measures. The total renewable electricity contribution above would represent about 40% of the EIA electricity projection without accounting for our efficiency improvements. This may seem high compared to relatively arbitrary goals, but it is consistent with what is economic and what is needed to mitigate climate change with renewables. If all these renewables were deployed together, because they would compete against each other, the total potential would be somewhat less than shown here. On the other hand, the various renewables occur in different regions, and a relatively conservative approach was taken to arrive at the each technologies contribution [100].

The map in Figure 4.40 shows how energy efficiency and the various renewables considered in the study could be distributed throughout the United States. Although there are some regional differences, there is a fairly even distribution of all these renewable energy options and energy efficiency across the country [100].

CSP uses direct solar radiation in desert regions to supply wholesale electricity and is fairly constant from early morning to the evening due to 6 hours of inexpensive thermal storage. It can also be augmented with biofuels or natural gas to improve dispatchability. PV on buildings uses total solar radiation in populated areas to provide electricity on the retail side of the meter and, with no storage, plays out in the later afternoon. Wind often provides greater energy at night than during the day and complements solar options. Biomass and geothermal provide baseload power. Biofuels, of course, compete against gasoline. This blending of renewables is a balanced approach to electric generation. As PHEVs are introduced in increasing numbers over the next 20 years, they will provide a large amount of electrical energy storage throughout the country that pays for itself by avoiding gasoline costs. This can be made available to the electric grid via a smart electric grid and balance the electric system. This balanced approach of renewable energy sources combined with energy efficiency leads to a fairly reasonable approach to energy overall.

There will be some interest in maintaining a diverse portfolio of renewable options aside from purely economic considerations, and we are already seeing this with many state renewable portfolio standards. The renewable electric production technologies each had limited grid penetrations, with wind being the highest at 20%. A better understanding via analysis of the dynamics of an integrated renewable energy mix is needed. Also needed are projections of shifting the huge transportation energy to nighttime charging of the pluggable hybrid-electric fleet that will start coming on line around 2010. This will allow greater penetration of wind energy into the grid especially if the conversion to a smart grid occurs on the same time frame.

These studies did not consider ocean power or much thermal energy from renewables. On-site solar industrial process heat and solar heating/cooling could potentially provide additional car-

TABLE 4.8: Annual renewable electricity percent of grid in 2030 (TW·h) (from Ref. [100])

TECHNOLOGY	POWER (GW)	ANNUAL ELECTRICITY (TW·h)	PERCENT OF GRID ENERGY
CSP	80	300	7
PVs	200	300	7
Wind	250	860	20
Biomass	36	355	8.3
Geothermal	40	395	9.2
Total	606	2208	51.5

bon reductions. Although the studies included 6-hour thermal storage for CSP, they did not include electrical storage (e.g., batteries for PV or adiabatic compressed air energy storage for wind). Also, the studies did not consider superconducting or HVDC transmission lines, which would allow geothermal, CSP, and wind power to be distributed more efficiently over larger distances. Finally, the various forms of ocean energy were not considered because there is currently very little work on these technologies in the United States. All of these could increase the carbon reduction potentials in 2030 above those estimated in this report.

4.2.8.2 Cost of Carbonless Energy Options. A recent study [101] evaluated these nine carbonless energy options to estimate the cost of enacting this strategy. The approach taken was to note the amount of energy generated or saved and calculate the cost of doing so between now and 2030.

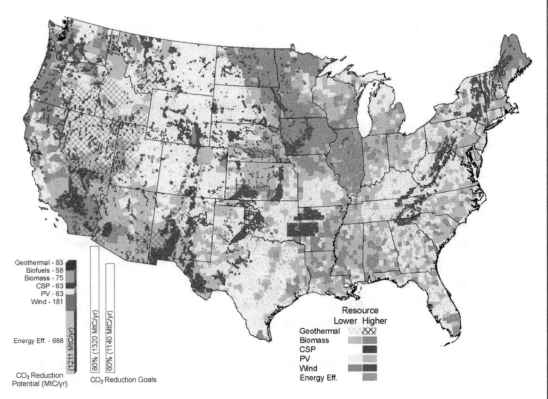

FIGURE 4.40: Distribution of the potential contributions by energy efficiency and renewable energy by 2030. (from Ref. [100])

The cost of the conventional energy displaced was estimated, and the difference between the two projections was the resulting net cost (or saving) of this carbonless strategy.

The deployment curve for each option in gigawatt-hours or gallons of ethanol or energy saved (oil, natural gas, or electricity) through efficiency from 2005 to 2030 was used based on the results in the prior sections in this chapter. The cost of each option was noted in each year that it was deployed, which considered learning curves used in the analysis. The total cost in equivalent $2005 was determined by using standard life-cycle cost analysis with an 8% discount rate. The project costs of conventional energy displace were determined using the EIA conservation (low) cost projections. The annual costs for deploying each of the nine options are shown in Table 4.9 [101].

The results show costs savings of $108 billion per year for all the energy efficiency techniques. The renewable options range from zero cost for wind to about $9 billion per year for cellulosic ethanol. All the six renewable options cost about $26 billion/year. The overall results of all nine options are that about $82 billion are saved each year. Even the $30 billion/year cost the six renewables option is not unreasonable to address a planetary crisis. Further consideration of the policy implications are given in Chapter 6.

There is no accounting in this analysis of the savings from the reduction in the cost of fossil fuels that would result from the deployment of this amount of energy efficiency and renewable energy. However, Office of Management and Budget examining the McCain–Lieberman bipartisan cap-and-trade legislation (S139) found that the economic results of this legislation tended to stabilize fossil fuel prices and accrue economic benefits to citizens. This amounted to an additional $48 billion annual by 2020. The magnitude of fossil fuel cost savings would more than pay for supports for the host of renewable energy options.

4.2.9   Green Jobs

The accelerated use of energy efficiency and renewable technology options identified in this book is referred to as the alternative energy strategy. Although it certainly is smaller than the "convention" energy sector (fossil and nuclear), what is surprising is the current magnitude of these alternatives. In 2006, these industries in the United States had nearly $1 trillion in gross revenues, created nearly 8.5 million jobs, and generated over $150 billion in federal, state, and local government tax revenues. Incidentally, they have already displaced a large amount of fossil energy as summarized in Table 4.10 [102].

In the renewable energy sectors (RE), the gross revenues were about $35 billion and the total jobs were over 400,000. The jobs created were disproportionately for scientific, technical, professional, and skilled workers. About 95% of the jobs were in private industry. Nearly 70% of

TABLE 4.9: Annual costs to deploy carbonless options (from Ref. [101])

CARBONLESS OPTION	COST ($B)
Energy efficiency	−108
Wind	0
Biofuels	9
Biomass	3
PV	5
CSP	7
Geothermal	2
Total	−82

the jobs were in the biomass sector—primarily ethanol and biomass power. The second largest number of jobs was in the wind sector of the industry, followed by the geothermal sectors. More than half of the RE jobs in government (federal, state, and local) were R&D-oriented jobs at DOE laboratories. The RE sector contains some of the most rapidly growing industries in the world, such as wind, PVs, fuel cells, and biofuels with long-term double-digit growth. The contribution from hydroelectric part of renewable energy sector is ignored since the potential for additional facilities are limited.

In the energy efficiency (EE) sector, gross revenues totaled $933 billion. The total number of jobs created by EE exceeded 8 million and 98% of the jobs were in private industry. More than 50% of the jobs were in the manufacturing sector and the second largest number of jobs was in recycling, followed by the construction industry. Nearly 80% of the EE government jobs were in state and local government [102].

Projections of the amount of EE and RE to 2030, predict that the range of revenues from these industries could vary from $1.9 trillion to $4.5 trillion and the total jobs (direct and indirect) would be from 16.3 million to 40.1 million (Table 4.11). The base case is essentially the EIA projection to 2030 where the amount of renewable electricity is about the same as today at 9% of total and the amount of ethanol increases from 4 to about 15 billion gallons This scenario ignores the climate change issue as well as limitations to the global supply of oil. The advanced case is similar to the

TABLE 4.10: Revenue and jobs in energy efficiency and renewable energy in 2006 (from Ref. [102])

INDUSTRY	REVENUES ($ BILLIONS)	DIRECT JOBS	TOTAL JOBS CREATED (DIRECT PLUS INDIRECT)
Renewable energy (1)	35.2	186,000	427,000
Energy efficiency	932.6	3,498,000	8,046,000
Total	967.8	3,684,000	8,473,000

From Bezdek (2007) [102].

(1) Considers private sector only and excludes government jobs related to renewable energy.

estimates of EE and RE contained in this chapter where about 50% of the electricity is provided by RE and about 60 billion gallons of ethanol are generated from various biomass sources.

Based on comparison to coal and nuclear power plants, the amount of labor involved in renewable energy is larger per unit of electric energy generated. Studies indicate that the amount of jobs involved is 5 to 10 times more for concentrating solar thermal electric power plant based on a life-cycle analysis. A similar story exists for renewables and energy efficiency measures compared to oil and natural gas energy [103]. Often the RE and EE jobs are closer to home because the energy efficiency and some forms of RE (distributed PV and solar thermal) are within the urban area that

TABLE 4.11: Revenue and jobs in energy efficiency and renewable energy in 2030

	REVENUES ($ BILLIONS)			TOTAL JOBS CREATED (DIRECT PLUS INDIRECT, IN THOUSANDS)		
	BASE	MODERATE	ADVANCED	BASE	MODERATE	ADVANCED
RE	95	227	597	1305	3138	7918
EE	1818	2152	3933	14,953	17,825	32,185
Total	1913	2379	4530	16,258	20,963	40,103

From Bezdek (2007) [102].

these services are used. Thus, not only are the jobs per unit of energy more intensive for EE and RE, but they are near the end use of this energy service.

Taken together, RE and EE reduce carbon emission risks, lower oil and natural gas prices, facilitate an industrial boom, create tens of millions of jobs, foster new technology, revitalize the manufacturing sector, enhance economic growth, and help eliminate the trade and budget deficits.

. . . .

CHAPTER 5

Conventional Energy

5.1   "CLEAN" COAL

Coal use in the United States is enormous. It is used to generate half of the U.S. electricity and is responsible for one third of the total carbon dioxide (CO_2) emitted. Coal is viewed by some as the bulwark of America's energy future. Its claimed outstanding attributes are that there are enormous amounts of it in the United States (George W. Bush publicly claimed that there is 500 million years of coal in an energy independence speech in 2005), that coal is cheap, using it is getting cleaner, and truly "clean" coal is just around the corner. It is also claimed that if we limit CO_2, the price of electricity will skyrocket and the economy will collapse. Is there any basis for these claims?

As it turns out, there is less than 100 years of coal accessible at today's price and consumption levels. Coal can be cleaner at the power plant and its pollution at about the level of a relatively clean natural gas plant. The resulting CO_2 can be sequestered in long-term geologic strata. However, even if national energy policy were implemented that enabled this very different use of coal, there is still the egregious impacts of gathering and transporting the coal. The cost increase to achieve this level of emission controls including CO_2 sequestration will not cause the ruination of the economy, and the cost of retail electricity would go up about 20% from about 10 to 12 cents/kW·h if all the coal plants were of this new "clean" type and it would take several decades to accomplish this transition.

However, there are two major issues:

1.   The stability of coal sequestration underground currently has unresolved issues that are site specific and would require certification of each depository to hold CO_2 for a millennium. This would delay siting of a new coal plant with CO_2 capture capability by at least a decade if an organization was established in the near term to certify long-term CO_2 repositories.
2.   The U.S. coal industry is embarking on a major power plant building episode. Together with other countries, if all these coal plants are built as planned between now and 2030, the amount of CO_2 released over their planned 60-year lifetimes will be the same as all the CO_2 released by humans since the beginning of the industrial revolution more than two centuries ago. If we are to achieve the necessary reduction in planetary CO_2 emissions, then these new coal plants must not be built as planned.

A strategy is needed now to allow the use of coal to meet near term and interim needs and yet not overwhelm the planet with a blanket of CO_2.

5.1.1   Tons of Coal

Each year, 1.16 billion tons of coals are mined in the United States. Ninety percent of this coal is used to generate half of our electricity, which is the last major use of coal in the United States. The remaining 10% goes to the chemical industry, which uses coal for raw feed stocks. Coal took over from wood in the 1800s as the major U.S. energy source, and its primary energy market share peaked in 1920 at 70% of all U.S. energy. Its market share has been dropping steadily since then and currently, the coal market share of U.S. primary energy is 23%.

Coal was eased out of most of its applications first by oil and then a combination of oil and natural gas. Price was not a driver in this huge transition over the last 80 years, and coal has been and still is still the lowest cost fuel on a Btu (unit of heat) basis. What impelled this transition away from coal was a combination of factors that could be characterized as energy system streamlining. Compared to oil and gas, coal is more difficult to gather, transport, distribute and use, and it has a large amount of solid wastes. It is also dirty and polluting, and its use has been moved out of the cities to more remote sites such as electric power plants. We do not use coal directly. We use the electricity make from coal, and it is the electricity that is much more streamlined in its system characteristics [104]. Incidentally, you can make the same observations about nuclear because it is the electricity from nuclear power plants that is streamlined and fits well into modern energy use in our urban center-dominated society. In addition, most renewable energy resources (concentrating solar power plants, geothermal, wind, and bioelectricity) use electricity as the energy carrier to the urban centers).

Is George W. Bush correct when he announced that we have 250 million years of coal? Is there an almost unlimited reservoir of coal in the United States (and China) that we would be fools not to use it to power our continued prosperity. Well, how much coal is there?

The current official U.S. estimate from the Energy Information Agency (EIA) of the Department of Energy (DOE) is that the United States has 270 billion tons (Bt) of coal, which is about 25% of the world reserve. We currently use 1.16 Bt of coal each year, so this would lead one to say that we have about 250 years of coal at our current rate of use. The official U.S. estimate has progressively come down from about 3000 Bt of coal in 1909 (Marius Campbell and Edward Parker of the USGS), to 483 Bt reserve base in 1974 (Paul Averitt of the USGS). Of this amount, only about half was considered accessible and recoverable. The current EIA estimate of 270 Bt is still based on this 1974 data, which itself was based on state geologist's data that is 50 to 60 years old.

It is interesting to note that there have been attempts to pry into this with more detail. The USGS Kentucky geologic survey in 1986 examined 60 square miles and found that only 33% was recoverable. When this was enlarged to 20 sections of 60 square miles each in Kentucky, West Virginia, and Virginia, it also found that less than 50% was recoverable. In 1989, Tim Rohrbacher of USGS introduced economics more explicitly and looked at the first 60-square mile section. He found that only 22% instead of 33% was recoverable at $30 per ton (2006 contract price for delivered coal was $32/ton). When he examined sections from Illinois and Colorado, he estimated that the recoverable amount of coal was between 5% and 20%. Even in the substantial Power River coal field in Wyoming, Rohrbacher found that only 11% was recoverable.

If these latest findings were applied to the EIA estimate, there would be about 68 Bt of coal that was recoverable. When you ask the sedimentary geologists Rich Bonskowski who is the person in charge of coal numbers at EIA (the only one in charge), he admits that the current EIA estimate bothers him. "I'd like to be able to bring sober assessment to those numbers, but we just don't have the resources" [105]. It does not look like he is going to get help anytime soon.

Every fossil fuel is limited and depleteable. Even coal has a relatively near-term production horizon of about 70 years in the United States at current price of about $30/ton. The current minor use of coal goes to the chemical industry as feedstock for a myriad of products and uses 10% of the current coal production. Coal as a chemical industry feedstock seems like the major long-term use of coal and one that should be saved for future generations.

5.1.2 How Dirty Is Dirty?

On a day-to-day basis, coal is by far the dirtiest and most dangerous power source known to human kind. It is filthy at every step of the way between getting it out of the ground to the voluminous solid and gaseous wastes that are left over after the electricity is generated. The filth goes hand in hand with the danger.

The mining operations have killed more than 100,000 miners in the United States in the last century and caused the deaths [106] of another 200,000 in a more lingering agony of black lung caused by coal dust (conservatively bases on data from Ref. [107]). Three quarters of all coal are moved by rail and this amounts to 60,000 coal trains each year that are 1 mile long. This huge caravan of rolling stock takes their toll of life at railroad crossings.

Coal plant stack exhaust pollutants kill 24,000 Americans each year and cause 400,000 severe asthma episodes [108]. This level of fatalities is more than those who died of murder, AIDS, or drug overdose. This obscene number of fatalities rivals the mayhem on American roads. However, you control your car and this gives the illusion that you are in control of your destiny on the road. You do not control the American coal industry. They control you. It almost seems normal that no death certificate has ever listed coal plant pollution as the cause of anyone's death in the United States.

Coal mining has moved increasingly from deep coal mining to mountain top removal and open pit mining to expose the coal. The benefit of this evolution is that the number of workers needed has lowered over the last four decades reducing the exposure to various forms of death while increasing the amount of coal removed. The price for this industry "improvement" is that they are regularly removing mountain tops.

Then, the question is, what do you do with the mountain top? In Appalachia, they just drop the mountain top into the nearest valley and have buried more than 1200 miles of rivers (Figure 5.1). This has caused the destruction of local water systems, and in general, wreaking havoc. Six thousand eight hundred toxic sludge impoundment ponds exist, and many of these are just uphill of small communities [109]. One third of these sludge ponds have failed in West Virginia alone [110]. Acid mine drainage from waste coal and abandoned mines leaks out and combine with groundwater and streams, causing water pollution and damaging soils. In Pennsylvania alone, acid mine drainage has polluted more than 3000 miles of streams and ground waters, which affects all four major river basins in the state [111]. The injured locals have no recourse but to endure their losses because Big Coal controls the levers of power in these Eastern mountain states [105].

Although coal produces only 23% of our total primary energy, coal power plants generate air pollution on a prodigious scale and produce 40% of the CO_2 in the United States, 67% of the sulfur dioxide, 22% of the nitrogen oxides, and 33% of the mercury plus 60 other pollutants including lead, chromium, lead, hydrogen chloride, and arsenic [112]. Acid rain has been an historic problem that has ruined thousands of lakes that only recently has some hope of reducing impacts (the extension of the Clean Air Act in 1990 to include a cap and trade agreement with some coal plant operators to control SO_2).

After mercury is released in the plant exhaust, it enters the air and is carried down by rain into our streams, lakes, and other waters where it poisons the fish and seafood that eventually make their way to our dinner tables. Mercury accumulates in fish and the animals and people who eat them, causing brain damage, mental retardation, and other developmental problems in unborn children and infants [113]. It has also been linked to a greater risk of coronary heart disease in men

FIGURE 5.1: Mountaintop removal coal mine in southern WV encroaching on a small community. (Photo by Vivian Stockman.)

[114]. As a result, 47 U.S. states and territories list mercury fish eating advisories. Mercury's most insidious effect is the reduction of the intelligence of a newborn, which placed them at a functional disadvantage for life.

Harmful air pollution is also released when coal is transported. About 75% of all coal shipments in the United States are made via railroads, which are one of the nation's largest sources of soot and smog pollution. Both soot and smog can cause health problems, including respiratory problems and increased risk of asthma attacks. Coal-laden railcars also cause soot pollution when coal dust blows off into the surrounding air, a substantial problem considering that a typical coal plant requires 40 railcars per day to deliver the 1.4 million tons of coal needed each year [115].

More and more rail coal movements bring cheaper coal from the West to East. Railroad congestion is the biggest bottleneck to expanding coal fired power plants. Although coal is not thought of as an energy source with vulnerabilities to terrorism, the removal of only one railroad bridge over the Mississippi would have disastrous impact on Eastern electricity production.

Beyond conventional air pollution and copious amounts of CO_2, coal mining is also a source of other greenhouse gases (GHGs). Methane, a major global warming gas, is about 25 times more potent as CO_2, is found trapped around seams of coal. It is released from the surrounding rocks when coal is mined, as well as during coal washing and transportation. Coal mining releases about 26% of all energy-related methane emissions in the United States each year [116].

The recent shift to natural gas-fired power plants is an attempt to avoid most of these impacts and the reluctance of the coal power industry to build new power plants after the Clear Air Act Extension was passed in 1970. Average emissions from a natural gas combined cycle (CC) power plant produces half as much CO_2, less than a third as much nitrogen oxides, and 1% as much sulfur oxides. It also avoids most of the problems in gathering and transporting coal, although newer gas fields in Montana do have significant local environmental impacts. The price volatility of natural gas and the prospect of CO_2 control have renewed interest in new coal plants before the end of the Bush era. However, the continued shift to natural gas away from coal would reap a significant reduction in pollution as well as carbon release.

The claim that coal electricity is "cheap" is so outrageous that words are difficult to formulate. The price we all pay for coal-generated electricity is significant. The license that we have all granted the coal industry to dump their wastes on top of the rest of us is almost without bounds. The United States attempted to bring coal pollution under some control with the historic Clean Air Act Extension and Amendments passed between 1970 and 1990. New sources of power were to perform as best as was possible using the best available technology. This law grandfathered existing coal plants because they were expected to be shut down at the end of their normal lives. As the new plants were built to be as clean as was reasonable and the old dirty plants were retired, we all expected the air to indeed be cleaner over time. This was to be done inexpensively because studies showed that for each dollar in pollution reduction cost, there was a $40 reduction in damages [105].

Since 1970, the coal industry sharply limited its construction of new coal plants and has extended the lives of the old polluting plants by essentially rebuilding them in place in the guise of performing "regular" maintenance. This flouting of the law of the land for more than 30 years borders on criminal because of the damage inflicted by not replacing these polluters in a timely manner. What makes this hard to understand was that this national law created a level playing field for all providers of coal power plants. These were reasonable rules for all reasonable businesses to use as a normal part of doing business. Yet the coal power plant industry chooses to frustrate the law and to continue polluting. This flies in the face of all that is decent, and even conservative economist Milton Friedman said that, "there is one and only one social responsibility of business and it is to use its resources and engage in activities designed to increase its profits so long as it stays within the rules of the game" [117].

Recently, the G.W. Bush administration relieved the coal power industry of the Clean Air Act standards and created the Clean Skies program, which uses a cap-and-trade approach to controlling coal plant pollution. Cap and trade was a successful type of approach developed in the 1990 extension of the Clean Air Act when directed at acid rain. The Bush Clean Skies is another matter. According to Goodell, "This program is not working because the cap is loose rather than firm, the governing rules are poorly designed rather than precise, the penalties for exceed the cap are low rather than significant, and the timetable for implementation is long rather than short" [105]. Even a good policy can be undercut by politicians who are impervious to the issue and beholding to their corporate supporters. This key issue will be discussed in Chapter 6 when politics and policy are considered.

5.1.3 New Coal Plant Designs

5.1.3.1 Integrated Gasified Combined Cycle. The claim that coal electricity is cheap is only accurate when we talk about internal costs, and we ignore the dreadful price we all pay by allowing the coal industry to get away with the mayhem that they generate throughout the land. Because we all pay most of the cost of coal electricity with injury, death, and environmental disruption, the marketplace cost of coal is indeed low.

The internal price of electricity from some old coal plants is about 3 cents/kW·h based on the average spot price of $3/MBtu (million Btu) in July, 2008 (EIA at http://www.eia.doe.gov/cneaf/coal/page/coalnews/coalmar.html#spot) with 0.75 cents/kW·h operation and maintenance costs for a subcritical pulverized coal plant at 34.4% efficiency. The actual fuel cost to a particular plant varies because of the nature of long-term contracts and the shipping costs to this plant. The old coal plant cost of about 3 cents/kW·h is quite low because the construction cost has been paid off decades ago and all the external costs are ignored. To understand the actual cost of the coal electric industry, one would have to catalogue all the impacts, disruptions, deaths, and injuries and give each a dollar value. Add these all up and divide by the energy generated by all the coal plants to get an estimate of the rest of the costs of coal electricity. This would mean we have to put a dollar value on a human life, injuries, the loss of IQ points in a child, planetary overheating, etc. This is a difficult and truly odious enterprise for a democracy where people are supposed to have some influence over their own lives through the democratic process.

Instead, what if we took the approach of estimating the cost of coal electricity by removing some of the egregious impacts of coal at the coal plant? As a starting point, consider a new (supercritical) and somewhat cleaner traditional coal plant costs about 7.5 cents/kW·h. A new integrated gasification CC (IGCC) technology coal plant that eliminates much of the air pollution (about 33% of the NO_x, 99% of the SO_x, 10% of the CO_2) would cost about 7.9 cents/kW·h (based on the Future of Coal, MIT, 2007, with current coal price of \$3/MBtu and with capital costs increased by 50% to reflect recent increase in construction materials). There is a 0.46 cents/kW·h cost increment to significantly clean up most of the pollutants except the CO_2 in a coal plants exhaust.

Because the gasification primarily produces carbon monoxide and hydrogen, the combustion products are CO_2 and water. The IGCC technology does give good access to the CO_2 by simply condensing the water and pressurizing the CO_2 to a supercritical liquid. Small amounts of remaining gases are vented out the stack. The CO_2 can then be pumped into long-term storage (sequestration) most probably in an underground saline aquifer. "Carbon capture has the potential to reduce more than 90% of an individual plant's carbon emissions," says Lynn Orr, director of Global Climate and Energy Project at Stanford University.

The extra cost of IGCC plus commercial sequestration technology is estimated to raise the power plant cost to about 10 cents/kW·h after pilot plants were demonstrated and the sequestration stability was assured. If you try to do this with a conventional coal plant, it would cost about 11.6 cents/kW·h because it is much harder to get at the CO_2 in the hot flue gases (Michael Mudd, CEO of American Electric Power).

Thus, for a new coal plant, the wholesale cost increment is about 6% for cleaning up much of the exhaust pollutants (about 0.46 cents/kW·h). When you include sequestering CO_2, the wholesale electricity cost produced by the plant goes up an additional 2.2 cents/kW·h or nearly 35% more than the electricity from a conventional new coal plant. How does this affect your retail electricity cost?

The average U.S. price of retail electricity is close to 9.1 cents/kW·h, and the average wholesale price is 6.2 [118]. About half of our current electricity is derived form coal. If all our coal plants were converted overnight to this IGCC type with CO_2 sequestration, the average U.S. retail price for electricity would increase from 9.1 to about 11.6 cents/kW·h, which is about a 27% increase. Of course, this would take many decades to accomplish and would probably be in the noise of general inflation. I will let you decide if this increase in electricity cost is worth the enormous reduction in health and environmental impacts including underground sequestering the largest single source of GHGs in the United States.

The IGCC part of this new coal plant is apparently a commercial reality. Of the more than 150 new coal plants that the coal industry has recently put forward for development, 30 would use the IGCC design. The gasification part of the IGCC plant is based on a 150-year-old technology where coal was gasified to make "coal gas" or "town gas," which was used to light cities before the electricity era. This same coal gas is currently used today in the chemical industry.

The CC part of the plant is simply the combustion of gas in a gas turbine-generator, and the hot exhaust gases then boil water making steam to spin a steam turbine-generator. This is the primary type of fossil power plant built in the last 30 years and a common commercial product. This type of electric-generating plant is nearly as clean a natural gas plant and much more efficient than the traditional coal plant (about 52% efficiency vs. 34–43%). IGCC is a relatively straightforward combination of these two prior commercial products to generate electricity at more than 50% improvement in efficiency and significantly less exhaust pollution.

The first coal IGCC demo plant (Cool Water, California) was built in 1984 and ran until 1989. In the 1990s, coal IGCC plants were built in Indiana, Florida, the Netherlands, and Spain. Using waste oil, IGCC plants are built worldwide and all are still in operation. All of this is relatively straightforward, and the proof of the readiness of IGCC plants is that 30 of the new crop of recently proposed coal plants were the IGCC technology.

5.1.3.2 Oxygen-Fired Pulverized Coal Combustion. There is a major alternative approach to the IGCC process that allows CO2 capture based more or less on conventional coal plant technology. This approach seems to have greater interest in Europe and it is called oxygen-fired pulverized coal combustion (oxy-fuel) process. It has similar cost characteristics of the IGCC process. The oxy-fuel process has the distinct advantage that it can be retrofitted on existing conventional pulverized fuel (PF) coal plants built at different level of coal plant efficiency (subcritical or super/ultrasuper critical).

Simply put, the oxy-firing of PF in boilers involves the combustion of pulverized coal in a mixture of oxygen and recirculated flue gas rather than ambient air. This reduces the net volume of flue gases and substantially increase the concentration of CO_2 in the flue gases (95%) compared to the normal pulverized coal combustion in air (15%). The high concentration of CO_2 can be compressed to a liquid, and any remaining nitrogen or argon separated and vented out the stack. Oxy-combustion is likely to give increased fuel flexibility, which would be about the same as current pulverized coal.

In addition to giving access to CO_2 in conventional coal plants, there would be significant reductions in the capital and operating cost of flue gas-cleaning equipment such as NOx reduction because oxygen is used instead of air to combust the coal. The SO_2 would be removed using conventional scrubbers.

There are a number of variants for the proposed oxy-firing of conventional PF boilers, but in simple terms the technology involves modification to familiar PF technology to add oxygen separation, flue gas recycling, and CO_2 compression, transport, and storage. A typical arrangement is shown in Figure 5.2.

Relatively pure oxygen is mixed with a proportion of either wet or dry flue gas taken downstream of the particulate cleaning plant (typically 70% of the total gas flow) and blown into the wind box of the boiler. Primary air to sweep the pulverizing mills is substituted with dry flue gas. The net result of this combustion process is a concentrated stream of CO_2 that enables the CO_2 to be captured in a more cost effective manner.

Technical challenges remain and include investigation of flame stability, heat transfer, level of flue gas clean-up necessary and acceptable level of nitrogen and other contaminants for CO_2 compression, and corrosion due to elevated concentrations of SO_2/SO_3 and H_2O in the flue gas.

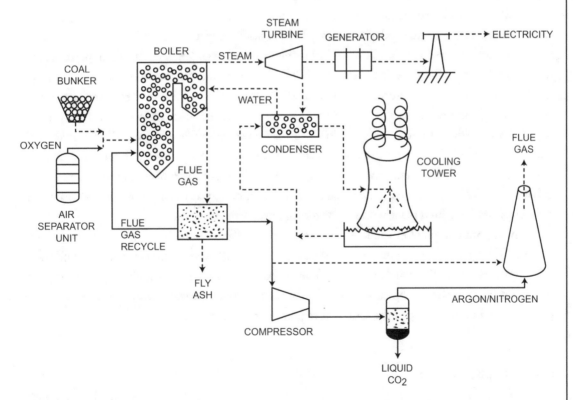

FIGURE 5.2: Oxy-firing process: oxygen combustion with recycled flue gas (dotted lines shows conventional pulverized fuel plant).

To date, pilot-scale studies have demonstrated that there are no significant technical barriers to oxygen and recycled flue gas firing of PF boilers. No full commercial scale demonstrations are being planned. However, a number of investigations have used pilot-scale facilities in the United States, Europe, Japan, and Canada.

Studies have also assessed the feasibility and economics of retrofits and new power plant. Findings from these studies indicate that preliminary cost evaluations indicate CO_2 capture costs and electricity costs are comparable with other technologies and less expensive than conventional PF coal plant with postcombustion capture of CO_2.

A working group composed up of Australian and Japanese organizations is currently engaged in a 2-year feasibility study and will probably retrofit a 30-MW boiler as the next step. Vattenfall, a Swedish company, broke ground in Germany in mid-2007 for a pilot coal-fired plant to demonstrate uses the oxy-fuel technology and will be the first large-scale test of whether carbon can be captured and stored using this method. This facility will completely integrate the process and take the CO_2 to a site where natural gas is being extracted, and pump the CO_2 into the ground to stimulate gas recovery. "It's a well-proven technology, which does not really need to be demonstrated, but Vattenfall wants the process to go full-cycle," according to Lars Strömberg, the Vattenfall COE [119].

Because of oxy-fuel's simplicity and its suitability for lower-grade coals, a U.S. power equipment manufacturer (Babcock & Wilcox) has been taking a strong interest in this approach. Speaking to a congressional committee in March, 2007, Babcock CEO John Fees said the company is putting its eggs in the oxy-fuel basket. In the summer of 2007, Babcock & Wilcox has been testing oxy-fuel combustion at a research plant in Ohio that is the same size as Vattenfall's German facility. However, the Ohio test facility does not have the equipment to separate oxygen and nitrogen on-site or to capture CO_2 and store it in the ground at the end of the process. According to Fees, it is Babcock & Wilcox's next corporate goal to use this technology as the first commercial-scale, near zero-emission, coal-fired plant with carbon storage in North America.

Advanced coal plant efficiencies have approach 44% and can be integrated with the oxy-fuel process. If designed for this integration, advantage can be taken to maximize efficiency and minimize cost such as smaller boiler size [120]. The overall efficiency needs to include the cryo-separation of oxygen, which will consume about 22% of the energy of the coal plant, as well as additional power needed to pressurize of the CO_2 before pumping it into the ground.

5.1.4   Sequestration

Taking things one step further would involve the sequestration of CO_2 to essentially remove the last major pollutant from the IGCC and oxy-fuel plants. As with coal gasification and CC power plants, sequestration is also an industrial process and it has primarily been used in the oil and gas industries. Called enhanced oil or gas recovery, the CO_2 is pumped and liquefied, and injected into the oil field to increase the productivity of an older oil field by much as 50% and has been used in Texas oil fields for 20 years.

The Weyburn oil field in Saskatchewan, Canada, is using CO_2 to augment production. The unique aspect of this facility is that the CO_2 comes from a gasification plant in Beulah, ND, and the CO_2 is pumped 205 miles to Weyburn (the CO_2 is a byproduct of the production of gasifying coal). Six million tons (Mt) of CO_2 have been injected into this oil field in the first 5 years. Six Mt of CO_2 have been sequestrated under the North Sea in depleted gas fields. A typical 700 700-MW coal plant uses 1.4 Mt coal/year/y which, which produces about 5.5 Mt CO_2/year. This rate of CO_2 emission is about five times5 times greater rate of CO_2 as is sequestered in the Weyburn facility. This is a similar order of magnitude and should be within commercial reach in early demonstration plants. Although enhanced oil recovery is a common practice which, which has not resulted in apparent environmental impact, geoscientists have not rigorously assessed how long the CO_2remains underground.

About 95% of all the locations where CO_2 can be injected into the ground in the United States are in deep saline aquifers that are several thousand feet below the surface. There is sufficient capacity to store all the CO_2 emitted by U.S. industrial sources, and the estimate capacity is 1100 GtC. There is about 520 GtC of this capacity that is available at a cost of less than $60/ton CO_2 ($220/ton carbon) to capture and store the CO_2 [121]. The current U.S. emissions of carbon is 1.6 GtC/year, which pails besides this capacity to store the carbon. "To move all the U.S. CO_2 underground would need an infrastructure that is similar in size to that of the current oil and gas industry," according to Lynn Orr, Director of Global Climate at Stanford University. Not surprisingly, it will take a huge infrastructure to completely deal with a huge problem.

There are a number of environmental questions such as the possible release of heavy metals, which might seep into drinking water somewhere above the deep saline aquifer. CO_2 makes water acidic, which can dissolve rocks and release minerals, which might include arsenic and uranium. These is a possible issue with leakage of the CO_2 itself. It is buoyant underground and may migrate and pool. At concentrations greater than 20%, CO_2 is an asphyxiant. We need to turn these "mights" and "mays" into a better understanding of how CO_2 behaves in deep saline aquifers before moving ahead rapidly with CO_2 sequestration.

The Frio experiments at the University of Texas at Austin have helped to answer the CO_2 stability question, at least for some types of geology. The October 2004 pilot project was the first test in the United States to monitor the capacity of the subsurface to store injected CO_2. Scientists successfully used a computer model to predict that injected CO_2 would move a short distance through pores in the rock and brine and quickly come to a stop. The pilot also demonstrated the ability of instruments to measure the movement and location of CO_2.

Based on this and other initial work, some researchers are quite confident that CO_2 sequestration will work and permanently store the CO_2 in saline aquifers. "It will never come out," says geologist Susan Hovorka of the University of Texas at Austin, who has been conducting carbon sequestration feasibility experiments. "It's moving through the tiny pores between the sand grains and it gets smeared, like grease on a tie." Hovorka's initial experiments at an oil field northeast of Houston have shown that the CO_2 behaves as expected, remaining trapped in the geologic formation [122]. Although this is encouraging, there should be some caution about pushing CO_2 sequestration too quickly. Clearly, a much more vigorous national-level research, design, and development (RD&D) program is required to give us the insight into CO_2 sequestration that is needed for large-scale commercial operations.

A recent example of support for sequestration research is that in October 2007, the Bureau of Economic Geology at the University of Texas at Austin received a 10-year, $38 million subcontract to conduct the first intensively monitored, long-term project in the United States studying the feasibility of injecting a large volume of CO_2 for underground storage. The DOE-funded research program is to demonstrate CO_2 injection rate and storage capacity in the Tuscaloosa–Woodbine geologic system that stretches from Texas to Florida. The project will inject CO_2 at the rate of one million tons per year, for up to 1.5 years, into brine up to 10,000 feet below the land surface near the Cranfield oil field about 15 miles east of Natchez, Mississippi. Experimental equipment will measure the ability of the subsurface to accept and retain CO_2. Sue Hovorka does not think acidification and mineral leaching will turn out to be a deal breaker. She is actually more concerned about displacement of salty water. As the CO_2 is pumped in at high pressure, some of the salty water is displaced, potentially contaminating fresher water above, which might be used for drinking. Or it might alter overlying ecosystems. This side effect might limit how much or how long carbon sequestration can be safely done in an area. The current crop of DOE programs, which are investigating these very questions, should be supported and rapidly accelerated.

Ed Rubin of Carnegie Mellon University, one of the authors of the special report to the IPCC on carbon capture and sequestration (CCS), says that we need about 10 full-scale demonstrations of CCS across the country. These need to done in different utility environments, using different types of coals, in different coal plants (both IGCC and oxy-fuel), and sequestering the CO_2 in different geologies. He suggests that this is the only way to answer the questions that utility companies have about a new process.

Good underground CO_2 storage areas are in Texan, Wyoming, and Illinois, and poor areas are in New England and the Carolinas. Internationally, the United States and Canada are better than Western Europe, which is better than Japan and Korea.

Other examples of carbon sequestration at a U.S. coal plant that have some promise can be found at utility company Luminant's pilot version at its Big Brown Steam Electric Station in

Fairfield, TX. This system is converting carbon from smokestacks into baking soda. Skyonic, who developed the process, plans to circumvent storage problems of liquid CO_2 by storing baking soda in mines, landfills, or simply to be sold as industrial or food grade baking soda. GreenFuel Technology Corporation is piloting and implementing algae-based carbon capture, circumventing storage issues by then converting algae into biofuel or feed.

Sequestration in geologic reservoirs—oil and gas reservoirs and aquifers—is much more promising, provided that the reservoirs will remain intact. The collective leak rates of the reservoirs must be significantly lower than 1%, sustained over a century-to-millennium-type time scale. Otherwise, after 50 to 100 years of sequestration, the yearly emissions will be comparable to the emission levels that were supposed to be mitigated in the first place. CO_2 geysers can form naturally and have been known to harm life when the CO_2 migrates out. Furthermore, because every underground geologic aquifer is different, being able to verify that billions of tons of CO_2 could be sequestered in one or two particular aquifers for centuries to millennia at a 0.1% per year leak level would not be sufficient to base an industry upon. We would need to perform such verification for every one of the hundreds of places that sequestration would take place.

Because the key to ongoing use of coal for electric generation is sequestration for the long term, the certification of a site for CO_2 sequestration is as serious an issue as the siting of a nuclear power plant. Because of the difficult safety issues in the nuclear industry, the Nuclear Regulatory Commission (NRC) exists to give needed oversight. Similarly, a national level of review is needed for CO_2 sequestration. The Carbon Sequestration Commission, or the equivalent, would give oversight and review of applications for approval of a site for long-term (millennium) interment of CO_2. Without this national-level approach, each state would be responsible for proper oversight to CO_2 sequestration. This is a poor idea given the influence the coal industry has demonstrated over some state political process. Even at the national level, there is a significant issue that the Carbon Sequestration Commission may not be beyond the reach of Big Coal.

The current Bush administration promise of clean coal someday is embodied in the Future-Gen program, which is a prototype 275-MW coal plant using the IGCC approach and involves an international public–private partnership with DOE paying 74% of the bill. It would produce hydrogen as well as electricity and sequester the CO_2 under a limestone formation. Announced in 2003 and planned to start in 2012, FutureGen was to demonstrate this technology for the first time near Mattoon in southeastern Illinois. Delays and cost overruns have caused the project to be shut down [123]. Even if successful, this single program was too little too late.

In addition to the dominant use of coal for electricity, there is some interest in a coal-to-transportation liquids option. This is an old technology that was used by the Nazi to support their World War II ambitions. Lacking a reliable oil source to power their war machine, in desperation they used coal to generate transportation liquids. It is expensive and inefficient and makes diesel fuel not gasoline. However, if the problem of GHGs is paramount, then coal to liquids is especially foolish. It generates twice the CO_2 per gallon as using petroleum directly. A significant attempt was made in the 1970s to develop a cost-effective and clean approach to coal-liquids and it failed after billions of dollars of investment. It will also have all the problems associated with digging and transporting coal. This idea should be dropped.

5.1.5 "Clean" Coal Strategy

There are about 70 years of coal accessible at today's price and consumption levels. Coal can be "acceptably" cleaner if pollution is limited to that of a new natural gas plant and the CO_2 is sequestered. However, even then, there are still the egregious impacts of gathering and transporting the coal. The cost increase to achieve this level of emission controls and sequestration seems reasonable when the avoided air pollution health impacts and the significant reduction in emissions are considered. This is especially true when the CO_2 reduction is considered in light of the role coal plays in global warming.

Currently, some countries are on a coal plant building binge. According to the International Energy Agency, new worldwide coal plants are projected to number about 1400 by 2030. About half of these are slated for China, 15% in India, and the rest in the West. The United States has about 150 new plants somewhere in the application process, although a number have already been aborted. If all these prospective plants are built, they will generate about 570 Bt CO_2 over their 60-year lifetime, which is the same as all the CO_2 generated worldwide over the last 250 years. Clearly, leadership is required, and it is required now if we are to avert a planetary disaster. The United States absolutely needs to take a leadership role and reverse its historical foot dragging posture on climate change actions.

The overuse of coal is one of the main culprits of global warming. Coal now carries the burden of generating half the U.S. electricity, and a strategy is needed to ease coal into a transitional role with careful attention to the CO_2 emissions. As pointed out in earlier chapters, renewables and energy efficiency hold promise toward moving us to our 2050 goal of reducing carbon emissions by about 80%. This approach would meet the EIA business-as-usual projection of energy growth by a combination of energy efficiency (growing energy services with less energy used) and six renewable energy technologies used in parallel. Because there would be no need for greater use of fossil fuels such as coal, there would be no need to build new coal plants to power increasing energy use. The first step in this national strategy would be to use the reduced energy demand to allow the historical shift to natural gas for efficient electricity generation to continue. This shift would allow us to start retiring old coal plants and reduce the net CO_2 by more than 50%. As long as the price of natural gas is relatively stable, the cost increment for this substitute for coal is modest.

One way to drive the economics so that efficient gas CC plants are built instead of a coal plant is for a carbon tax of \$60–100 per ton of carbon [124]. This initial approach reduces CO_2 emission immediately to about half of burning coal directly. Other steps must be taken to go beyond this 50% CO_2 reduction.

In parallel with this initial step would be the certification of long-term sequestration sites around the United States. The necessary next step is the multiple full-scale demonstration of new coal plants with CO_2 sequestration as recommend by Ed Rubin. These new coal plants with carbon sequestration could be built to more rapidly retire the old coal fleet and this would relieve the pressure on gas supply. If the oxy-fuel technology pans out, it would also be possible to use it to retrofit some of the more efficient existing coal plants rather than retire them. This would only be attractive if these existing coal plants are situated near good sequestration sites whether saline aquifers, or oil or gas fields. Using this as an overall strategy, it is possible for coal to be a transitional fuel over the next century with significantly reduced CO_2 emissions.

It clearly does not make any sense to build up to 150 new coal plants in the United States as a last gasp of an archaic industry before the inevitable changes that are coming. A revenue neutral carbon dumping fee and/or a carbon emission cap-and-trade system are clearly in our national interests, and one or the other or both will be adopted in the near term. A new coal plant that does not consider the emerging carbon regulatory environment is indeed a risky project. "It would be important not to grandfather any existing or new coal plants," says Granger Morgan the head of Carnegie Mellon's Department of Engineering and Public Policy in an article in *Science* magazine. The last attempt at allowing grandfathering of existing coal plants as part of the Clean Air legislation of the 1970s backfired badly for the American people. We cannot afford to make this mistake again.

Recognizing this, Citigroup, JPMorgan Chase, and Morgan Stanley from Wall Street have announced that they have adopted a new set of environmental standards based on carbon emission to help lenders evaluate risks associated with investments in coal-fired power plants [125].

This leads to an important issue about this seemingly straightforward approach to "clean" coal. This strategy does cost more internally at the plant, and the extensive external benefits go to all of us. Why is this a problem if Americans are quite willing to pay 27% more for retail electricity on average over the next few decades? The Peabody Energy (the largest private-sector coal company in the world) and the other Big Coal companies would not get the benefit. They would have to do something differently and the reason for the change is not in their self-interest. In a political–economic system such as that in the United States, where these large private companies have extraordinary leverage over the political process, very little happens if it is not in their immediate self-interest. This situation where a greater public good would be achieved without particularly benefiting the private section is not a difficult issue for many countries, but it is for the United States.

Recognizing the realities of the current situation, there are some executives from the coal industry that have broken ranks such as James Roger, CEO of Duke Energy; Paul Anderson, chair of Duke Energy Board; and Wayne Brunetti, CEO Xcel Energy. These and other individuals support industry changes that would support reducing GHGs such as CO_2. Jim Rogers says his company will not build any more coal-fired power generators unless they have the potential to capture and sequester carbon emissions. Rogers said carbon sequestration, a process that stores CO_2 underground, will not work in the Carolinas, so the utility is only considering future coal-fired plants elsewhere. This decision reflects the growing sentiment that the United States should be actively fighting global warming, not building new coal plants that will make the problem worse. States such as Kansas and Florida have already said no to new coal, focusing instead on cleaner energy solutions such as wind, solar, and energy efficiency [126]. This issue of dominate corporate interests interfering with the general good will be considered in the next chapter.

A strategy is suggested that sees coal as an interim fuel over the next century at whose end; coal will be used primarily for chemical feed stocks. The "clean coal" strategy recommends that national policy be put in place that:

- In the near term, encourages efficient combined cycle natural gas plants that are relatively clean to replace older coal plants and to be used instead of new traditional coal plants. This type of gas plant generates less than half the CO_2 of today's coal plants at a cost similar to a new coal plant. (Rapid introduction of energy efficiency and renewables should stabilize the price of natural gas.)
- Increases RD&D for CO_2 sequestration near existing coal plants and near good transmission connections.
- Establishes a national Carbon Sequestration Commission to certify sites for long-term CO_2 impounding and give oversight to the national sequestration program.
- Demonstrates the use of the newer technologies (IGCC and oxy-fuel) at full scale with CO_2 sequestration in different geologic environments using different types of coal.
- Encourages efficient existing coal plants that happen to be located in a good CO_2 sequestration area to be retrofitted with the oxy-fuel process, which allows clean operation and the separation of CO_2.
- Encourages new IGCC and oxy-fuel coal plants with carbon capture and sequestration to be built to finally retire all the remaining old coal plants.
- Congress should remove any expectation that exiting coal plants and new construction without CO_2 capture will be "grandfathered" and granted emission allowances in the event of future regulation.

To expedite this program, a well-designed cap-and-trade mechanism or a revenue neutral carbon dumping fee of at least \$42/ton of CO_2 be applied to the power generation industry. If the revenue neutral fee is used, it would place a fee on a bad activity (dumping CO_2 into the atmosphere) and reduce taxes on personal income, investment gains, and other positive activities. This will provide the economic incentive to move away from coal combustion with uncontrolled CO_2 dumping in the atmosphere. The alternative is an effective cap-and-trade process, and this will be discussed in the next chapter.

The migration to carbon capture and sequestration coal plants would take about one to two decades to initiate. The actually implementation of building new "clean" coal plants would only be in the national interest only if energy efficiency (EE) and renewable energy (RE) options developed over this one to two decade time frame proved to be insufficient to meet our needs, or the goal of 80% carbon reduction proves inadequate. This "clean" coal option would be a kind of a very expensive insurance policy to have available on an interim basis if the reasonable projects for EE and RE in Chapter 4 actually turn out to not so reasonable.

Chapter 6 will consider policy approaches to achieving this strategy. Implementing this strategy for the continued use of coal for generating electricity would meet many but certainly not all of the characteristics of a "clean" energy system. Even if a coal plant was as clean as a gas fossil plant and emitted about 10% of the historical CO_2, it would still be cursed by all the impacts of digging out and transporting the black rock.

5.2 "ACCEPTABLE" NUCLEAR

Nuclear power in the United States has been dormant for a quarter of a century. The major issues leading to this shelving of new nuclear power were cost, environment and health risk, inability to establish a stable long (geologic)-term storage of highly radioactive wastes, and the connection to nuclear weapons of some fuel cycles. More recently, the ability of the NRC to manage the industry with sufficient independence of commercial concerns has come under question. The single most significant characteristic of nuclear power that is encouraging its reexamination as an energy option is that it does not emit CO_2 during operation.

An upbeat recent study by a group of professors from MIT [127] suggests that nuclear power could be one option for reducing carbon emissions and dealing with global warming because they do not emit carbon while operating. The MIT team realized that this is unlikely because nuclear power faces stagnation and decline. Recognizing the magnitude of the global warming threat to the planet, they tried to identify what would be required to retain nuclear power as a significant option for reducing GHG emissions. Their analysis was guided by a global growth scenario that would expand current worldwide nuclear-generating capacity almost threefold to 1000-MW reactors by the year 2050. Such a deployment would avoid 1.8 billion tons of carbon expected in a business-as-usual scenario. They wisely recommend that renewable energy sources, sequestration of carbon from coal plants, and increased energy efficiency be considered along with nuclear power but reached no conclusions about priorities among these options. They thought it would be a mistake to exclude any of these four options at this time. Given the historical curtailed role of nuclear energy in the United States and almost all of Europe, should nuclear be considered a major option, and what would it take to be "acceptable"? As a minimum, for the nuclear industry to be revived in the United States, it needs to deal with this cauldron of difficult issues.

5.2.1   How Expensive Is Expensive?

The last nuclear power plant that was built in the United States was ordered in 1978; it costs $7 billion (in the year of construction dollars) to build and was finished in the early 90s. More than 100 plants have been canceled since 1972, some when they were more than 40% complete. The current array of 104 nuclear power plants generates about 8% of our energy (20% of our electricity). The demise of this powerful industry was as much due to economic factors as to the negative response of environmentalists due to a number of unique risks that nuclear power presents. The final fiscal and environmental straw was the destruction of the nuclear reactor at Three Mile Island (TMI) near Harrisburg, PA, in 1979 because of the incompetence of the plant operators. In 90 minutes, NRC-trained and certified operators were able to make a $400 million investment into a $1 billion heap of radioactive rubble. "The nuclear recession was well along before TMI. What TMI did is further erode the confidence of the financial community in nuclear power,'" explained Dr. Shelby Brewer, the Department of Energy's assistant secretary for nuclear power.

The single most significant example of the magnitude of the economic disaster of nuclear power was the Washington Public Power Supply System (WPPSS) power cooperative that almost single handedly ruined the U.S. bond market by 1983. Although a number of factors drove costs overruns, the biggest cause of delays and overruns were mismanagement of the process by the directors and the managers of the system had no experience in projects of this scale. Of the five plants that were started, plants number 1, 3, 4, and 5 were never completed. Eight to ten billion dollars were invested and only one 1120-MW nuclear power plant was actually built. WPPSS defaulted on $2.25 billion in bonds, the largest default in the history of municipal finance [128].

A footnote on the way to nuclear stagnation was the Shoreham Plant on Long Island that cost $6 billion to complete and never generated 1 kW·h of electricity. Similar stories come from Europe where most counties (except France) decided in the 1970s and 1980s to no longer look to nuclear power to solve their significant energy problems.

The last time the United States had an active nuclear industry, the cost of U.S. nuclear plants escalated in cost by 10% per year above and beyond general industrial inflation [129]. This hyperinflation does not include power plants that were not finished even after significant investment. This means that the cost of a new nuclear power plant doubled every 7 years even as there was more experience building reactors. Instead of a learning curve to reduce costs of succeeding construction, there was a "forgetting curve" in the nuclear industry.

The 2005 Energy Policy Act also attempts to remove the financial risk to new power plant construction by having the federal government underwrite 80% of the financing. This is an enormous subsidy and attempts to override the market place in judging the risk of such a power plant. This end run around the market place putting a price tag on risks via financing conditions is also significantly moderated by the recent continuation of the Price Anderson Act. This act limits liability of a nuclear "incidents" to $300 million by the plant owner. Additional damage up to about $10 billion is paid from an industry fund where each nuclear plant-owning company pays about $96 million toward the "incident" at the rate of $15 million per year. Any damage above this is paid by the U.S. government.

This limiting the liability of the plant owner has a perverse impact on the safety of reactors in this country. If a plant runs for 50 years, it generates gross sales of electricity of about $20 billion. Let us say you believe that the chances of a major "event" is 1 in 20,000 each year. Over the life of the plant, you expect that there is only a 0.5% chance that you will have major liability and pay up to $300 million compared to sales of $20 billion. In this situation, most businesses would not pay too much attention to safety. They would run the plant into the ground and only do what was necessary to keep the plant on-line and bet on the 99.5% chance that they never have a major liability bill.

It would be wise to change this calculus so that the plant owners take safety more seriously because it will hurt their bank account not to. Raising the Price Anderson Act dictated liability limits from $300 million to $2 billion for your plant and from the $96 million to $500 million as your contribution for someone else's plant's "incident" should help the industry pay more attention to safety. This would change the relationship with the NRC from one that is adversarial to one that is cooperative.

A way to keep control of nuclear plant cost, construction schedule, and NRC safety updates would be to have a limited number of designs for all nuclear power plants—a cookie-cutter approach. Another approach is to use one construction team organization so that there would be a "learning" curve. With the current American practice of bidding and construction, this approach of one design and one construction team would not normally happen. The only way to achieve this significant cost saving for a nuclear power plant would be to have a monopoly that is government supervised such as in France. The monopoly could be open to rebid every 5 years to keep the contractor team more cost conscious. Another approach to controlling cost is the DOE 2010 initiative to reduce costs through new design certification, site banking, and combined construction and operation licenses. To the extent that this short-circuits reasonable review, it is a bad idea. The way this was set up in 2005 Energy Policy Act, it is a bad idea.

5.2.2   How Risky Is Risky?

The unique environmental risks of the nuclear industry are the magnitude of destruction and death from a low-probability nuclear core meltdown (caused by accident, incompetence, or terrorist act), and the difficulty in assuring that the dangerous radioactive waste products from the nuclear power plant can be isolated from our biosphere for geologic periods of at least 1,000,000 years. There is also a clear relationship between some nuclear fuel cycles and nuclear weapon-usable [the Pu from reprocessing is so called "reactor-grade," but can still be used to make nuclear weapons] materials that complicates the use of a nuclear reactor and introduce proliferation issues. This problem is being amplified because the United States is currently encouraging nuclear fuel cycles that will create weapon-usable material via the Global Nuclear Energy Partnership (GNEP) [134].

Most of the societal opposition to nuclear comes from these unique characteristics. A major stumbling block is the small probability of a core meltdown with the likelihood of a catastrophic number of resulting deaths and extensive damage. We do not seem to mind (so much) the fact that coal electric plants currently kills about 24,000 Americans a year [108]. This is a catastrophic number of fatalities but it happens over 365 days of the year and distributed over large regions of the country. Another example is that we accept 35,000 deaths a year from spatially and temporally dispersed vehicle accidents and from 30,000 deaths from hand guns.

However, we are horrified if 200 people die in one plane crash. With a nuclear power plant, a large loss of lives could happen at one location over days and weeks (acute deaths), and over years and decades for many more victims.

How large a loss of life and how big could the economic damages be? A recent study by Ed Lyman estimated the damage from a terrorist attack on the Indian Point nuclear plant located about 25 miles from New York City up the Hudson River. Using the same analytical approach done by the NRC itself, he concluded that it could result in up to 44,000 immediate deaths from radiation poisoning, 500,000 extra cancer deaths over the decades, and economic damages exceeding $2 trillion [130].

The reactor core meltdown issue is due to the exclusive characteristic of nuclear reactors that even after you shut the reactor off, it is not off. It still generates heat—about 3–5% of the full power heat. This is more than enough in the traditional reactor designs to exceed containment material melting temperature if active cooling is interrupted. With the containment compromised, the radioactive material and gases are at liberty to migrate from the reactor site causing extensive damage to the environs.

Although NYC has the greatest population near a nuclear reactor, there are more than 10 nuclear plant sites had more than 100,000 people living within a 10-mile radius. Understandably, society has difficulties with this characteristic of a technical system that essentially just boils water. Lyman limited his consideration to only the reactor and not the on-site spent fuel rods in open pools of water that contain much more radioactivity than the reactor.

Interestingly, the chance of a reactor core meltdown is calculated by the plant owners themselves using a computer model developed for this purpose [131]. The results of the owners exercising this computer software are that a core meltdown can occur every 20,000 years on average. This amounts to about 0.5% per year because there are about 100 reactors in the United States. Based on a 50-year life, the odds of a particular reactor melting down is about 0.25%. This ignores external events such as fires, earthquakes, accidents during shutdown, etc. These events would roughly increase the chances of failure by a factor of two [132]. This still does not consider the newly emerged higher possibility in the post-9/11 era of a terrorist attack. Without consideration of terrorism, there is about a 1% chance each year of 1 of the 100 U.S. reactors failing catastrophically.

If the U.S. nuclear reactor fleet is doubled or tripled, the NRC would have to increase the safety of nuclear reactors by a factor of two or three just to keep things as safe (or dangerous) as they are now. Unfortunately, the NRC is refusing to reconsider its long-standing policy of not requiring new reactors to be noticeably safer than today's aging fleet. This seems like a "head in the sand" policy. Is it necessary to have a major catastrophe to force its reasonable adjustment?

The remaining "external" event is a terrorist attack, and this would reduce the expectation of a severe "event" to less than once ever 100 years. How much less than a hundred years? We do not know, but the NRC says it is too low a probability to consider. NRC bases this judgment on their belief that terrorist acts are too remote and speculative. This does sound like the FBI and CIA attitudes about the possibility of significant terrorist's events before 9/11.

Even after 9/11, the NRC gives less consideration to terrorism than to plant accident or malfunction. A National Academy of Sciences study concluded that the near-term potential is high for civilian nuclear power plants to suffer from a ground or air assault as sophisticated as the 9/11 attacks. A large release of radioactivity could be one of the results [133].

Several problems stand in the way of addressing the risks of reactor sabotage and attack. The NRC gives less consideration to attacks and deliberate acts of sabotage than it does to accidents; the methodology for determining credible threats to nuclear facilities is flawed; and the process for determining whether reactor operators and the federal government can defend against such threats is inadequate [134].

This nuclear plant core meltdown can happen due to operator incompetence (examples are TMI in the United States and Chernobyl in the USSR), or in our current political situation, as an act of terrorism. Terrorism against a nuclear reactor could take the form of a "mole" nuclear plant operator activating a core meltdown purposefully. It could also be achieved by a small squad (15–20) of terrorists carrying hand weapons and taking over the operation of a nuclear plant. In tests of nuclear plant security, such small groups succeed in taking over the plant in more than 50% of the test trials. The results of such tests are no longer available to the public [135]. The private organization, Wackenhut, that currently provides about 30% of the plant protection is the same organization that is hired to check the adequacy of this security. This simple fact gives insight by those who are not technically informed, that there are major shortcomings with the current way that the nuclear industry is being run and receives oversight. A core meltdown could also be triggered by a follow-up to the 9/11 terrorist technique by flying a large plane into a nuclear plant.

Even in the wake of the 9/11 attacks, the NRC has blankly refused to consider terrorism in environmental impact studies on the grounds that terrorist acts are too distant a possibility. These EIS studies normally apply to a number of situations such as permitting for new reactors, 20-year extensions to operating licenses for nuclear plants, and licenses for expanding on-site spent fuel storage site [134].

In addition to not insisting new reactors be more safe to maintain even current safety levels, and not considering terrorism against a reactor, the NRC also does not consider the risk of terrorism against spent fuel rod water pools at the reactor site. Because this waste fuel releases heat and radiation, it needs to be stored about 5 years in water pools for years before it can be moved to permanent storage. The water cools the heat of the waste fuel but protects the site personnel from radiation. Draining the pool for even a few hours or stopping the active water cooling system for a few days would cause the fuel cladding to ignite spontaneously in air and melt the waste fuel. Because the pools typically contain about five times more fuel as the reactor, this cladding ignition would release very large amounts of radiation.

Again, the National Academy of Sciences reviewed the on-site waste fuel rods in open pools of water situation and concluded that a terrorist attack could cause the release of dangerous level of radiation under some conditions. They also concluded that the NRC does not fully understand the risks of this situation at every reactor site [136]. An attack against the waste fuel pool could result in thousands of deaths and billions in losses [137]. These results compare favorably with a Brookhaven National Laboratory study on stored waste fuel damage [138].

If the risks of terrorism are similar in magnitude to the other risks discussed earlier, then there is about a 2% chance each year of 1 of the 100 U.S. reactors failing catastrophically. It is difficult to know if this is conservative or not because this risk has been accentually ignored by the NRC. It is clear that the NRC should make a number of changes to reflect the current realities in the world. They should treat the risks of sabotage and attacks on par with the risks of nuclear accidents and require all environmental reviews during licensing to consider such threats. In addition, the NRC should require and test emergency plans for defending against severe acts of sabotage and terrorist attacks as well as accidents. The NRC should require that spent fuel at reactorsites be moved from storage pools to dry casks when it has cooled enough to do so (within 6 years) and that the dry casks be protected by earthen or gravel ramparts to minimize their vulnerability to terrorist attack. Current security standards are inadequate to defend against credible threats, and it is clear that an independent check is needed. Congress should give the responsibility for identifying credible threats and ensuring that security is adequate to the Department of Homeland Security [134].

5.2.3 New Reactor Designs to the Rescue

There are two ways to deal with the dangers of a core meltdown. Current designs can be improved and security can be improved with all the related difficulties of constantly assuring system integrity, or new designs should be pursued that make a core meltdown essentially difficult or impossible.[1] Both design approaches are underway, and there are a large number of new reactor designs. They are grouped in different categories, which roughly indicate their state of readiness. Generation III has three new reactor designs, which have been certified by NRC, and these are the Advanced Boiling Water Reactor (General Electric), the System 80+ (Westinghouse), and AP600 (Westinghouse). These focus on evolutionary refinements and are aimed at being safer and less costly. The Advanced Boiling Water Reactor (ABWR) and the System 80+ are quite close to existing designs. The AP600 attempts to reduce cost by "eliminating equipment subject to regulation" [139]. The AP600 does introduce some passive features, tries to eliminate some safety-related system, and reduces the size of the containment vessel. The net result of this attempt was that the reactor is neither safer nor less expensive. To overcome the cost issue, Westinghouse grew the size of the reactor and developed the AP1000.

Seven reactor designs are lumped into a category called Generation III+, and these are AP1000 (Westinghouse), Economically Simplified Boiling Water Reactor (General Electric), Evolutionary Power Reactor (EPR; Areva, France/Germany), Pebble Bed Modular Reactor (PBMR; Eskon, South Africa), U.S. Advanced Pressurized-Water Reactor (Mitsubishi, Japan), International Reactor Innovative and Secure (Westinghouse lead in international consortium), and the Super Safe, Small and Simple (Toshiba). Of these, only AP1000 is NRC-certified and the next four designs have applied for an NRC review. The AP1000 and the Economically Simplified Boiling Water Reactor both use passive safety features and a higher power rating to reduce cost. The 1700 MW U.S. Advanced Pressurized-Water Reactor (APWR) is a large evolutionary variant of today's pressurized-water reactors. Like the ABWR, it offers some incremental improvements over its Generation III counterparts, but it does not have novel features.

---

[1]See Nuclear Power in a Warming World for a fuller discussion of these issues.

The EPR is the only one that looks like it is appreciably safer than current reactors of all those new designs that are being looked at critically in the United States according to Lisbeth Gronlund. Known in Europe as the European Power Reactor, it was developed by a French-German team. It exceeds the NRC standards because it was designed to a more stringent safety criteria that was developed by the join team. It is being commercially pursued by Areva, which is a French company. It has a number of significant improvements over current U.S. designs such as a double-walled containment structure. There are four safety trains that are independent and each has a complete menu of safety systems to mitigate an accident. In addition, if the reactor core overheats and melts through the reactor vessel, there are systems to stabilize the core [134].

The EPR currently has precertification status in the United States. The first EPR, under construction at Olkiluoto, Finland, is scheduled to start operations by 2009. The second, in Flamanville, France, was approved by Electricite de France in May, 2007 and is scheduled for completion in 2012 [134].

As discussed earlier, the NRC should require new reactors built in the United States to be significantly safer than current reactors. To ensure that any new nuclear plants are significantly safer than existing ones, the NRC should require that new reactors have features designed to both prevent severe accidents and to mitigate them if they occur. These design features should reduce reliance on operator interventions in the event of an accident, which are inherently less dependable than built-in measures. This means that the NRC should not reward reactor providers who reduce safety margins no matter what PR phrases are attached to design changers, which do not improve safety. Even in the case of a safer reactor, the EPR may not be able to compete in the United States because the NRC is using unacceptably low safety standards, which place better designs at a disadvantage.

The other Generation III+ reactor of interest is the PBMR by Eskom. The PBMR represents another attempt to reduce capital costs through a design intended to be safer. The power plant would be made up of cookie-cutter modules that are used in tandem to make a large power plant. The module is sized at 150 MW of electric output. The use of repetitive smaller models could bring cost under control. The use of a smaller reactor as the building unit of the power plant could give greater protection in loss-of-coolant accident if the materials have higher temperature capability. The PBMR has all these characteristics and is inherently different from today's commercial light-water reactors. It uses helium gas as a coolant, a graphite moderator, and fuel consisting of very small uranium-oxide spheres coated with a corrosion-resistant material and embedded in tennis ball-sized graphite balls.

PBMR promoters claim the reactor as "inherently safe," arguing that the reactor's low power density and the high-temperature integrity of its fuel would prevent significant fuel damage, even in an accident in which the reactor lost all coolant (If the fuel retains its integrity, there is no radioactive release). This approach seems very promising and the design has entered the U.S. system.

After the U.S. utility Exelon withdrew the PBMR design from the NRC for preapplication review in 2002, the Pebble Bed Modular Reactor Co. (PBMR Ltd.), a consortium that includes British Nuclear Fuels and Eskom, informed the NRC in 2004 that it wanted to resume the preapplication review and intends to apply for design certification in 2007. In July 2006, Eskom submitted several white papers to the NRC as part of the preapplication review process.

The Super Safe, Small, and Simple reactor is a liquid sodium-cooled fast reactor that would provide 10 MWe of power and has a core lifetime of 30 years. A unit was offered by Toshiba to the town of Galena, AK, as a demonstration project. Toshiba has not yet initiated an NRC preapplication review.

A number of the Generation III+ reactors are fairly unique and beyond the professional experience of most of the NRC staff. This is a serious challenge for NRC, which has yet to overcome this particular issue.

Finally, there are five more advanced designs under development called Generation IV, which are the Very High Temperature Reactor, the Super-Critical-Water-Cooled Reactor, the Gas-Cooled Fast Reactor, the Lead-Cooled Fast Reactor, and the Sodium-Cooled Fast Reactor. The Department of Energy is sponsoring RD&D on these advanced reactor systems at national laboratories and universities. One of the goals of these advanced reactors is greater safety. However, there is no basis for assuming that any of the five designs now under study would be significantly safer than today's nuclear power plants. There is little operating experience for these approaches. For safety estimates to be realistic, there would be a need for detailed computer software to be developed to stimulate the operation of these new designs.

Under operating conditions, all the coolants used in these designs are highly corrosive. The containment for these fluids is advanced materials that need to operate in severe environments.

To reduce costs, Generation IV designs aim to reduce safety margins wherever possible. This is counter to the whole approach taken to reactor safety with a number of layers of protection with each being as independent as possible. This approach has been the cornerstone of reactor safety and mainly needed to make up for the vagaries of the primary safety mechanisms. Part of this safety strategy is to avoid the need for off-site emergency response plans. This aspect of safety is difficult to include in these advanced designs for a least a decade to gain operational experience. Kind of a catch 22 for these designs.

In turn, these reactor coolant fluids each have special problems. The Sodium-cooled Fast Reactor and Lead-cooled Fast Reactor have inherent safety problems because of their coolants. Lead–bismuth coolant is less reactive and has a higher boiling point than sodium coolant. However, it is extremely corrosive, and when irradiated produces highly volatile radioisotopes (polonium 210 in particular) that would be a challenge to contain even under normal operating conditions. In addition, if sodium boils, the nuclear reaction would intensify generating more heat and lead to an overtemperature condition. This is called a positive void coefficient of reactivity. This would rapidly lead to a thermal run-away and destruction of the reactor.

Beyond even these difficulties is the key design feature of these reactors—they are fast reactors. Up to this point, all commercial reactors in use or considered up to Generation III+ have been light-water thermal reactors. The fast reactors are different in one important respect. If the core should become more compact (meltdown), it would significantly increase reactivity. This would lead to a rapid power surge that could vaporize the fuel and blow up the core [140]. This would be similar to the explosion of a very small nuclear fission weapon and disperse the radioactive material widely. This characteristic of fast reactors would make it a hard sell for a commercial power reactor [134].

5.2.4   How Long Is Long Enough?

The need to insure geologically long-term interment of nuclear wastes is a significant risk of nuclear power. Currently, nuclear power plant wastes are located in water pools at the reactor site awaiting transport to the long-term waste facility. Without a containment vessel, they present a risk because of their vulnerability to terrorist threat. However, the current attempt to move these wastes to a long-term repository at Yucca Mountain, NV, is badly flawed and would not make things better.

The EPA design standard for the facility is that the storage site is to prevent leakage and seepage of waste for at least 1,000,000 years. It would take this long for the radioactive intensity of the nuclear wastes to decline to the level so that they can safely enter into the biosphere. Yucca Mountain has a number of issues like the salt deposit site in Lyons, Kansas, did before it. It is a volcanic remnant composed of pumice or layers of volcanic ash with a complex geology and is located in an active earthquake zone. A volcanic event may lead to magma intrusion into the site where the waste is to be stored, melting the metal canisters. This magma event could open a path to the surface releasing the radioactivity.

Water has migrated recently into the potential depository as noted by the presence of radioactive chlorine 36 in water found deep inside the mountain. This isotope could only have come from atmospheric nuclear testing in the 1950s and 1960s, which means that water has penetrated the mountain in fewer than 50 years. This is thousands of times faster than estimated by DOE [141]. This movement of water through the site would violate the 1,000,000-year containment period.

The temperature of the waste fuel canisters will eventually exceed the melting temperature of the rock that the tunnels, walls, and floor of the depository are made of. This will compromise the structural integrity of the depository. To overcome the melting of the rock, DOE came up with a scheme where a titanium "drip shield" would be placed over the canisters by remote control some hundred years from now just before the repository is finally closed. With the supporting rock melting, this scheme is unlikely to work.

Although these are major difficulties in meeting the geologic containment criteria at this site, EPA and the NRC did not decide to look for another site. They did decide to change the guidelines limiting the period of regulatory compliance to 10,000 years [142]. The Federal Court of Appeals (DC Circuit) overruled this reduction in storage time and said that the period of regulatory compliance had to extend to the peak dose time when the canisters would lose their integrity and start leaking. The repository itself had to contain the melted nuclear wastes after 10,000 years. EPA under the Bush administration then issued new regulations that elevated the allowable dose that could be released so that DOE would pass the necessary qualifications.

This truly is an example of voodoo science and political expedience dictating the numbers, not the science. So EPA cooked the books to bypass the Court of Appeals decision. It is clear that Yucca Mountain is not a proper site for depositing our nuclear wastes. It is also clear that we have a basic problem when politicians without much concern for environment and health risks for future generations can manipulate the science to suit their whims.

The MIT team dealt with this unacceptable situation by recommending that in the near term, a system of central facilities be established to store spent fuel in dry canisters for many decades. The spent fuel could then be moved to a new geologic disposal when it is properly certified and prepared. In parallel, the DOE should broaden its long-term waste R&D program to include improved engineered barriers, investigation of alternative geologic environments and sites, including deep bore hole disposal. This temporary storage facility of spent fuel placed in dry casks is economically viable for the next 50 years and can be secure if hardened against attack. Yucca Mountain should be abandoned from further consideration and another more suitable site should be found.

5.2.5 Making Electricity Generation Into a Nuclear Weapon

The current attempt by DOE to have a "closed" nuclear fuel system with weapon-grade materials placed in commercial circulation is one more major risk. The advantage of the "closed" fuel system is that it does not require any new nuclear materials to be mined. It would involve reprocessing spent nuclear fuels from today's burner reactors into a form for use in a new type of fast breeder reactor. The major drawback of this closed system would be that weapon-grade materials are generated and need to be transported around the country or world. This would be vulnerable to terrorist interception and use for destructive purposes. It also would complicate peaceful use of nuclear power in politically unstable countries. Finally, it would not improve the nuclear wastes problem.

To further this move toward a closed fuel system, the United States is trying to build a GNEP that attempts to safely use fuel reprocessing in future nuclear power plants. They claim that this approach can be proliferation-free and reduce nuclear waste issues. The MIT study team [143] does not see any benefits from this approach that would reduce nuclear waste problems. They feel that there is no need to generate weapon-grade material and to reprocess nuclear wastes. With a time horizon of the next 50 years, they recommend that the best choice to meet these challenges is the open once-through fuel cycle that we now use. They judge that there are adequate uranium resources available at reasonable cost to support this choice even under a global growth scenario that more than triples the amount of nuclear power in the world. Further analysis supports this finding and also finds that the GNEP approach would increase the risks of nuclear proliferation and terrorism [134]. The GNEP should be dropped as should fast reactors that use weapon-grade materials.

It makes eminent sense that the United States should reinstate its ban on nuclear fuel reprocessing, and the United States should take the lead in forging an indefinite global moratorium on reprocessing. The administration should also pursue a regime to place all uranium enrichment facilities under international control.

5.2.6 The NRC as Risk

A serious nuclear power accident has not occurred in the United States since 1979, when the TMI reactor in Pennsylvania experienced a partial core meltdown. Comforting as this is, it does not necessarily demonstrate that safety measures and NRC oversight are adequate.

To examine this possibility, a recent study found that individual reactors had to be shut down to restore safety standards 35 times since 1979. In each of these cases, the plant owner had to take more than a year to address a large number of equipment impairments that piled up over years. These are failures that should have been caught in a timely way but were not. That is, until the malfunctions reached the point where this extended downtime was needed to set things right.

Taking over a year to deal with the accumulated impairments shows that a serious problem exists with NRC oversight. An effective regulator should be neither unaware nor passive in the face of such extensive malfunctions [134].

One of these 35 instances occurred in 2002. A discovery was made that the Davis-Besse reactor in Ohio had a sizable hole in its top plate with a veneer of stainless steel to keep the core's radioactive materials contained. This would have led to an "incident" as bad or worse than the TMI core meltdown in a few months. Four dozen "abnormal occurrences" such as this were reported to Congress by the NRC since 1986.

What happened at the Davis-Besse reactor in Ohio and how does this episode reflect on the NRC's ability to give oversight to the nuclear industry and put the public at risk? In fall 2001, NRC staff members analyzed conditions at Davis-Besse and were so concerned that they drafted an order requiring the reactor to shut down for immediate inspection. They were concerned about parts they suspected were cracked and potentially leaking radioactive reactor cooling water. Cracks and leaks had been discovered in similar nuclear plants, such as Oconee in South Carolina. The NRC staff determined that Davis-Besse was highly vulnerable to the same degradation. So far, so good.

The NRC managers actually agreed with its staff that a disaster was eminent. In fact, the NRC determined that Davis-Besse was not meeting all five of the basic safety principles [144]. In spite of this, the NRC chose to put aside these known safety issues and bowed to requests of the plant owner to keep the reactor running until the next scheduled maintenance.

When the reactor was finally inspected months later, it was found that the NRC had indeed broken all five of its safety principals [145]. A large hole had been corroded through the head plate of the reactor vessel. Less than a quarter inch of steel was left in the many-inches thick head plate of the reactor vessel. The release of radioactive cooling water would have been disastrous because the two key backup systems were not working well enough to handle the emergency.

Researchers at Oak Ridge National Laboratory later evaluated how close Davis-Besse had come to disaster and concluded that the reactor head would have ruptured in 150 to 200 days [146]. Luck, not regulatory prowess, prevented that disaster from occurring [134].

This episode was especially troubling because a stronger case may never be assembled by the staff. The fact that the NRC agreed with its staff and determined that all the basic safety rules were being violated at Davis-Besse and still acting not as the safety overseer but as a flunky for the power plant owner. If the NRC could not or would not act in this case, then it is hard to imagine what type of a case it would act in. The NRC has lost its capability to give independent safety oversight.

There are a number of important changes that must be made before seriously considering expanding the nuclear industry in the United States. The most vital change is that a serious effort must be made to change the organizational culture of the Nuclear Regulatory Commission (NRC) and restore its safety oversight capability. Without overcoming the current NRC dysfunctionality, the remaining policy initiatives would be meaningless as would any attempt to restart the nuclear industry in the United States.

A serious question is how to change the core culture of an organization? In the private sector, this is usually taken care of by the eventual failure of the entity. In the public sector, this handy mechanism is not available. Techniques like a reorganization can be attempted but is unlikely to make any significant change. The NRC staff appears to be functioning relatively well, but it is the commissioners themselves along with their upper management who are not functioning properly.

The process of organization change must be driven from both the top and the bottom. The key usually is that organizational leaders must be identified. They are the individuals who are seen as leaders and respected by their co-workers. They are not necessarily the people with hierarchical leadership. These leaders need to buy-in to the new safety-first culture.

For example, the failure of GM to evolve its culture so that training someone to work on the assembly line would be like the more successful Toyota approach even after several decades indicates the difficulty of attempting organization change. The Los Angeles Police Department has been about to evolve its core culture from "enforcing the law" to "protecting the community". This was done by allowing the officers responsible for the communities to voice what was needed rather than a command-and-control process based on "this is what will be done" coming down from the top.

How do you change of the NRC core culture to focus on industry safety as its core vision? How do you turn loose this internal revival by empowering leaders throughout the organization? The commissioners are appointed by the president with Senate approval for 5-year terms. Currently, there is no requirement that the NRC commissions are safety experts and independent of nuclear industry manipulation. Clearly, replacing most or all of the commissioners as they come up on their 5-year terms is vital. To restructure the commissions would mean that Congress and the President would have to be on the same page to stimulate this turnover. The oversight committees of Congress would have to take a direct hand to stimulating the renewal. This is a difficult problem but one that must be solved before any serious thoughts of significantly increasing the size of the U.S. nuclear industry.

Rather than making it easier to build nuclear reactors by underwriting effective oversight, and as a minimum, Congress should require the NRC to bring in managers from outside the agency to rectify this problem" [134].

Contributing to this finding is the recent history of the NRC where the right of the public to engage in the reactor licensing process has been curtailed. This blocks an important means to improve safety because public input has long played a key role in the NRC deliberations. This has been recognized by the NRC itself in many instances. In spite of this, the NRC has dropped the public's right to discovery and cross-examination when either renewing power plant licenses or applying for a new license.

Nuclear is currently a dormant energy option in this county and most of the industrialized world and has proven to be relatively unattractive to almost all developed countries. It is relied on for significant percentage of power in only four industrialized countries (France, South Korea, Taiwan, and Japan) that share one key social/political characteristic. These countries all make power plant decisions in a central agency with little or no public input. Currently, China and India are now also intent on building new nuclear reactors. The 2005 Energy Policy Act as well as the NRC internal processes do have provisions that would sharply curtail and some say it would eliminate public involvement in the siting and approving future nuclear power plants. If this law is maintained in the next administration, it would attempt to make the United States similar to the six counties that exclude an open political process in evaluating the reasonableness proposed nuclear plants.

Finally, the NRC's budget is inadequate. Congress continues to pressure the NRC to cut its budget, so it spends fewer resources on overseeing safety. The NRC does not have enough funding to fulfill its current mandate to ensure safety. It certainly will not be able to adequately respond to new applications as well as to extend the licenses of existing reactors. Without an active oversight role by the NRC, it is difficult to imagine enlarging our fleet of nuclear reactors.

5.2.7   What Is "Acceptable"?

It is clear that nuclear reactors have a potential to reduce carbon emissions compared to fossil power plants. Given the difficulties facing us of reducing our global carbon footprint, nuclear should be given serious consideration as an option. What is not clear is if the cost and risk of using nuclear is manageable or acceptable. Rather than making it easier to build nuclear reactors by underwriting the cost of reactors with 80% financing guarantees, and almost eliminating the monetary damage of "incidents" with the cheap insurance of the Price–Anderson Act, the opposite should happen. Rather than coddling the industry and isolating it from the realities of the real world, it is time to take seriously the risk that this energy system poses. The industry was shut down last century by the financial market for all the right reasons. It was too expensive and too risky. This was not a failure. It was the market working correctly. To try and override this by suspending good judgment is not in the public interest. Let the market decide if we can afford the cost and the risk of starting up the nuclear industry again. Remove the unwarranted financial props to underwrite a fully commercial industry. Adjust the industry cost for "incidents" via the Price–Anderson Act to catch the plant operator's attention and give them an incentive to pay close attention to operational safety (increase liability from $300 million to $2 billion for your plant, and from the $96 million to $500 million as your contribution for someone else's plant's "incident").

What else is needed for an "acceptable" nuclear industry? If the NRC insisted that the industry come up with a safer reactor design such as the EPR; if we can remove the vulnerable radioactive wastes in water pools from reactor sites and put them in off-site dry casts and protect them; if the NRC was directed to take the terrorist threat seriously all these years after 9/11 and include it in the environmental impact proceedings; if the task of protecting reactors and these waste products were put in the hands of the Department of Homeland Security; if DOE would dump the Yucca dump site and move on to a suitable long-term repository; if the NRC would reintroduce public access and involvement in review of licensing procedures; if the provision in the 2005 Energy Policy Act that allow new design certification, site banking, and combined construction and operation licensing were dropped; if the NRC funding would increase so that the NRC had the resources to deal with their task; if we can break the link between boiling water and creating weapon materials and drop the GNEP; if the United States would reinstate its ban on nuclear fuel reprocessing; if the United States would support the open once-through fuel cycle that we now use; if the United States would take the lead in forging an indefinite global moratorium on reprocessing; if the United States would pursue a regime to place all uranium enrichment facilities under international control.

The final and probably the most difficult and crucial of the above scenarios is fostering change in the core culture of the NRC. Can the NRC change itself from a culture of looking the other way to please the nuclear industry financial interests to a culture where safety actually came first? This type of cultural change of an institution has been especially difficult to manage. I know of few successful examples. Without instituting this change, we are really rolling the dice when it comes to assuring the safety of this industry in the United States.

To achieve the institutional changes in the NRC so that it could properly lead the revitalization of the U.S. nuclear industry would take a minimum of 5 years (the time to replace all the sitting commissioners) and probably more to carry out the rest of the organizational revitalization. Five years for this first step assumes that the president and the Congress act on this policy recommendation and substantially increase the funding of the NRC. Special Congressional oversight would be needed to revamp the commission.

When the NRC is functioning again and has included terrorism as a factor in establishing safety designs and has increased the safety design requirements to go along with a much larger nuclear industry, the nuclear industry would have to respond with safer designs. This would take another 5 years as a minimum. This might be shortened somewhat if the nuclear industry actually took this policy seriously immediately and did not invest in attempts to politically manipulate this policy's implementation.

These new designs would have to be evaluated using extensive software simulation since there would be no operational experiences with these new designs. The revamped and geared-up NRC would need to develop the capability to give proper overview to these new designs before any of these propose designs could be approved. These three activities would take about 15 years to put in place with proper budgetary support and Congressional oversight. Reactors take from 6 to 12 years to complete and this would take us to about 2030 for the first batch of new reactors to come on line. This assumes that the cost of these new designs would have to be such that they are cost-competitive with other ways to produce no-carbon energy. That is, nuclear would have to be competitive with increasing energy efficiency and a host of renewable energy options. The cost of these new nuclear reactors would not be known until the nuclear industry makes the recommended changes.

The historic rate of building reactors in the United States was about 100 reactors were built in about 30 years. Even doubling this prior record, about 200 new reactors would be built by about 2050. These would replace the current set of reactors that would be retired by then, and this amounts to a net gain of 100 reactors over today. This amounts to about 200 GW of nuclear power and they would generate about 1550 TW·h of electricity. Assuming that the nuclear electricity would replace carbon emissions at an average rate of 210 t C/GW·h, these 200 new reactors would displace 330 million tons of carbon per year. This would be a significant contribution to carbon reduction and amounts to about 20% of today's total U.S. carbon emission.

By implementing this 13 policy option package, the new safer and functioning nuclear industry would be ready to play a major role by about 2025. At this time, we will have clear data on the magnitude and cost of the energy efficiency and other noncarbon (renewable) approaches to providing U.S. energy needs. If these noncarbon options efforts prove less than adequate or the severity of the climate warming problem proves greater than anticipated, this "acceptable" nuclear industry can be pushed with vigor to complement efficiency and renewables.

If all these ifs were accomplished, then we could and should embrace "acceptable" nuclear power. Time is short. It is time to move forward on all these ifs.

.   .   .   .

CHAPTER 6

Policy for Whom?

This study shows that a carbon-free opportunity exists to meet the challenge of reducing U.S. carbon emissions by 2030 while being cost competitive with today's energy costs. Some renewable options can generate wholesale electricity at 4 to 10 cents/kW·h, whereas others cost less than retail electricity cost for on-site systems in certain regions. Energy efficiency options are less than $2.30 to $5/MBtu natural gas (today's price is about $10/MBtu), less than $29/bbl oil (today's price is 50 to 150 $/bbl), or less than 4 cents/kW·h (today's price is closer to 10 cents/kW·h). The historical cost reductions of renewable options were considered to continue and future costs of conventional fuels and electricity were considered to increase but at a rate based on the very conservative (low) Energy Information Agency (EIA) projections. An example of EIA's conservativeness is their projection of the price of gasoline in 2030 to be $2.45/gallon. In this study, economically competitive models were often used and renewables had to compete with these very low EIA cost projections. Why bother to use EIA projections? These are the official U.S. government projections to be used by all government agencies. In spite of this assumption used throughout this study, the results show that carbon emissions can be reduced on a trajectory through 2030 to meet the 2050 goal of about 80% reduction compared to today. In addition, the overall cost savings by 2030 for the energy efficiency (EE) and renewable energy (RE) put in place in this scenario is at least $82 billion per year [101].

Since this projection of energy efficiency and renewables is cost competitive, there should be no extra cost incurred in the market place to reduce carbon emissions if this projection were to take place. Indeed, there should be direct cost savings without counting the reduction in the price of conventional fuels when the projected amounts of energy efficiency measures are introduced. There is some public investment over the next two decades for government research, design, and development (RD&D) and subsidies that are phased out by 2020 to help with the transition to meeting the carbon reduction goal by 2030.

This reduction of carbon emissions was accomplished by a combination of nine techniques based on energy efficiency and renewable energy options. A number of important options such as "clean" coal and "acceptable" nuclear were not part of this total. Nuclear and coal can play a role, but they are not acceptable in their current forms. However, these conventional energy options can be modified to contribute toward this carbon reduction goal using carbon sequestration for coal (see Section 5.1) and a host of achievable changes for nuclear (see Section 5.2). The measures needed to achieve these vastly improved coal and nuclear systems will make these systems available in about a decade or two. The "acceptable" nuclear and "clean" coal options could be activated if the gains projected for energy efficiency and renewable energy do not prove adequate or if the needed carbon reductions are greater than now thought. Although the initial goal of preventing carbon levels exceeding 450 to 550 ppm appears to inadequate. More current work indicates that we should keep global CO_2 level stabilized below 400 ppm [147]. Automotive options such as the near-term pluggable hybrid electric technology and the longer-term use of hydrogen as an energy carrier are also ignored in this time frame till 2030. Energy conservation was not considered as part of these projections although this option can make a powerful contribution to the future reduction of carbon emissions. The potential of carbon sequestration in forests and other flora or changes in agriculture practices in the United States were not included, although there are many opportunities to use these options to remove carbon from the atmosphere if done on a long-term sustainable basis. These opportunities can be examined on other day.

6.1   TILTED PLAYING FIELD

The outcome of this study with minor exceptions was based on the energy efficiency and renewable options being cost competitive with conventional energy sources. All nine of these options were considered in a conservative manner and only used to the extent that they were attractive in limited regions of the country without modifications to our electric transmission system or the introduction of commercial-scale energy storage systems such as cost-effective underground compressed air systems. All in all, these results are conservative in almost all ways except one. It is assumed that all these decisions to use cost-effective energy efficiency and renewables took place on a level playing field not biased by the current influence of conventional energy system operators either politically or economically. This study assumes that the millions of people making countless decisions each do so based on economic self-interest and behave perfectly rationally.

If the desirable outcome of significantly reduced carbon emission can be achieved at very competitive economics, why are any policy changes needed for this outcome to be achieved? The basic reason is that there is not a level playing field for renewables and energy efficiency at this time.

The current energy system exists today as a result of a century of development, investment, and a physical, economic, and political structure created to accommodate the current conventional energy system. Power lines and pipelines are where they are now not to accommodate the vast potential of various renewable energy sources. There are located where they are to accommodate yesterday's conventional energy sources. The electricity pricing rate structure of today's utilities was not developed to their current form to accommodate user-owned on-site energy options. Examples of on-site systems are photovoltaics (PV) on buildings or commercial combined heat and power

(CHP) systems that would compete with retail electricity and fuels within the urban center. Current government RD&D subsidies were not designed to support energy options that are new and promising, and are to be the energy system of the future. Current government RD&D props up energy options that have a large industrial base of support and have been commercial energy systems for more than 50 years such as oil, gas, coal, and nuclear. Even the most recent National Energy Plan (2005) only allocates 5% of the plan's support to renewables.

An enormous amount of energy efficiency in all sectors of our economy are very cost effective, but there is little or no interest in taking advantage of these cost savings over the operational life of a particular system. Buildings have a 30- to 100-year lifetime, yet essentially zero percent of these buildings are designed to be energy efficient over their actual lifetime. Owners of industrial processes are overly first cost sensitive and corporations routinely reject 2-year breakeven improvements in process efficiency. Yes, American industrialists routinely reject process modifications with a 40% rate of return. So much for economic rationality.

Buildings almost always are built by people who do not live in them, so there is little incentive for the builder to worry too much about energy operating expenses. Even energy-intensive commercial diesel trucks are built to be energy cost effective over their first 3 years, although this equipment is used for 15 years. The reason is simple. The first owner typically uses the truck for 3 years. This type of list can go, and it would take a shelf of books to capture most of the examples.

So the good news is the transition to increased energy efficiency and renewables will be driven by the increasingly competitive energy economics [59]. The bad news is that the lack of a level playing field will delay the introduction and use of renewables as well as energy efficiency. Because of the delays that will result from this tilted situation, the time scale for this transition is out of step with the timing of the needed action on avoiding the worse impacts of climate change. Policy is needed to make the inevitable happen sooner rather than later when it will not do us much good. Why do I say that the delay won't do us much good? To meet an 80% CO_2 reduction works if we start in 2010. If were start in 2020, it will require an 8% per year reduction in CO_2 [147].

New rules are needed for all energy stake holders to use so that we capture the power of the market forces and yet avoid the worse impacts of global climate change. These new rules (policy) need to recognize the built-in structural disconnects between the old and new energy system, as well as the economically irrational decisions that are commonly made throughout our energy economy. New rules are needed to recognize the damage inflicted on all of us by carbon and other greenhouse gas (GHG) emissions that are not considered in current energy economics.

6.2 UNIQUE DIFFICULTIES WITH CLIMATE CHANGE

In addition, there are at least two serious sets of issues that make this seemingly straightforward task on introducing cost-effective energy efficiency and renewable energy difficult and bordering on impossible. The first is due to the issues related to the unique aspects of climate change—it takes a long time to see the results of the carbon and other GHG overload of the atmosphere. Without seeing the negative results, it is hard to mobilize the needed political reactions to negate the problem. We are dependent on long-term scientific projections as the basis for our near-term mitigating actions. Using scientific studies as the basis for a political decision is particularly difficult for the United States. It is much more than U.S. scientific illiteracy illustrated by the example that only 26% of Americans think that evolution was caused by natural selection [148].

Also, our political system and its decision-making mechanisms are based on a relatively short-term time horizon of 2 to 8 years at most. Yet carbon dioxide (CO_2) residence time in the atmosphere is from 50 to 150 years. This is a large mismatch between the time scale of the problem and the time scale of our political horizon. Yet somehow, some political decisions supporting a particular project have had a much longer life. An example is the Interstate (Eisenhower) Highway System in the United States took about 35 years to complete and cost $425 billion in 2006 dollars. Another is the state of California's support for a bevy of measures to stabilize electricity consumption over the last 30 years while California has had a rather politically extreme sequence of governors.

There is some scientific uncertainty over the cause and effects of climate change although the uncertainty is decreasing over time. The current Intergovernmental Panel on Climate Change report [149] judges that there is a 90% probability that the current warming is human caused. Many people would think that 9 chances out of 10 are pretty good odds, but even that small uncertainty is big enough to drive a Hummer through if your world view normally does not recognize pollution as a problem. Also, global warming is caused by so many different human activities taking place all over the world in differing magnitudes and over different time scales of 150 years. This tends to get people pointing fingers at the other and makes any negotiation based on who is responsible and equity issues extremely difficult.

Finally, there is the North–South stalemate with the governments of the South pointing out that the Northern countries have thus far been responsible for the overwhelming majority of the GHG release (the United States, Europe, and Japan have contributed 70% of the total CO_2 dumped into the atmosphere to date). The South will be impacted more than the North and it is poorer with fewer resources to adapt. So the South argues that it would be unfair to require them to make any immediate contributions to preventing global warming. The North should go first and then help the South acquire climate-friendly technology. Meanwhile, the North points out that any reductions in their emissions would be rendered ineffectual if emerging countries such as Brazil, India, and China did not curb the growth of their GHG emissions.

Additionally, there is the ethical issue of differential damage. Projections of impacts (especially drought and water issues) will be especially severe on developing countries such as sub-Sahara Africa) and yet the developed countries caused most of the damaging GHG. How do you deal with this disparity?

6.3 A START

To get the policy ball rolling, let us look at a recent economic study by Nicholas Stern et al. called the Stern Report. This was a result of a call for an independent review by the Chancellor of the Exchequer reporting to both the Chancellor and the Prime Minister of the UK [150]. The bottom line was that from the many economic perspectives used by Stern, "the evidence gathered by the Review leads to a simple conclusion: the benefits of strong, early action (mitigating GHG emissions) considerably outweigh the costs." The Report goes on the explain that, "the evidence shows that ignoring climate change will eventually damage economic growth. Our actions over the coming few decades could create risks of major disruption to economic and social activity, later in this century and in the next, on a scale similar to those associated with the great wars and the economic depression of the first half of the 20th century. And it will be difficult or impossible to reverse these changes. Tackling climate change is the pro-growth strategy for the longer term, and it can be done in a way that does not cap the aspirations for growth of rich or poor countries. The earlier effective action is taken, the less costly it will be."

The numbers from the Stern Report concludes that the annual cost of stabilizing the CO_2 level at around 525 ppm will be about 1% GDP by 2050 and further concludes that this is a significant but manageable level. Doing nothing, the Stern Report concluded, would cause damage from climate change that would cost the world between 5% and 20% of the world's gross domestic product.

Stern notes that most other study results to deal with climate change cluster between −2% and +5% of GDP by 2050 and says this is due to uncertainty and key assumptions such as the discount rate. Most of the criticism of the Stern report by economists seems to center on the discount rate assumption. The discount rate is the device used by economists to compare the economic value of something (usually money) today with tomorrow. It is a common feature of money market analysis to value different investments.

For example, Stern uses 1.4% for the discount rate and this translates into a trillion dollars in 2050 being worth $497 billion today. William Nordhaus looked at the same issues and used a 6% discount rate, which means that a trillion dollars in 2050 is worth $50 billion today. Quite a different result. The economists who attacked Stern accused him of making an ethical choice by using a relatively low discount rate. That is, he gave more consideration to future generations than a 6% discount rate would imply. They insisted that economists should not introduce ethical values into such calculations. In this case, what is the value of an investment today to prevent damages to future generations. Yet they insisted on using the 6% discount rate and said this was not injecting ethics into the analysis. They used 6% because it is commonly used by economists dealing with money markets. They seem satisfied that they were being ethically neutral with this stance. I wonder?

Yes, they are making an ethical value judgment as choosing any discount rate would imply. The question is, if it is okay to use a 6% discount rate, which is normally used to put future and present money on similar footing. Is it okay to use this value when trying to consider today's dollars compared to tomorrows damages including loss of human life and species extinctions? This certainly is an ethical question of the highest order, and the economists who took issue with Stern need to do more than hide behind the discount rate used in money market analysis.

However, Stern's conclusion that "the benefits of strong, early action (mitigating GHG emissions) considerably outweigh the costs" does get us off to a running start in the attempt to formulate policy as a reasonable response to the monumental issue of dealing with human caused climate change.

6.4   ENERGY TRIBES

The second broad set of issues has to do with the makeup of our political beliefs system that governs all of our political discourse. There seems to be different sets of beliefs that are separate and not equal with almost no overlapping areas of agreement [151, 152]. This makes for difficult political decision making in the best of circumstances.

You may have noticed that some people believe that there are no environmental problems and that these pseudo-problems are simply opportunities. They think that all science on climate change is junk science (Senator James Inhofe (R) of Oklahoma). They believe that the only credible solutions to energy issues are developing more supply of existing energy forms and to let market forces dominate the outcomes. They also believe that the ablest should get the most rewards. Yet others feel that we cannot take a step forward without legion of experts who will guide our every step. Others yet, think, feel, and believe that humankind are flawed and the only solution is to withdrawal from engaging in any activity that has environmental impacts of any type. Furthermore, they believe that all corporations act in a crazed manner to generate growth in products, in profits, and in endless impacts. They perceive that this is the problem, and doing without is the solution. And there is an undercurrent of folks who do not believe any problem is solvable and people are beyond redemption. These are four sets of views, each dramatically different. It is important to note that we step into and out of these different forms of cultural values in different parts of our lives and sometimes often on a daily basis. These internally consistent holistic views of the world and coping strategies are not personality types [153].

I am sure you have met some of this cast of characters. How could you miss them, they are us! How do you form a rational approach to dealing with climate change when these are the people who make up who we are—when there are so many nonoverlapping sets of rationality? What common ground is there is move toward to fashion a solution to the difficult problem of human generated climate change?

The major question that must be answered before a coherent policy can be drafted to address climate change is, "policy for whom?"

If I suggest policy for all those that agree with my basic contentions, it would be a feel-good exercise for the subset of Americans that agree with me, but what about the rest? My policy would harden the resolve of some and drive them on to overturn the political influence of all these ill-advised folks. Others would say, do not waste your time with these futile attempts to solve this problem. Yet others would accuse me of selling out to the enemy if I offer any accommodation of other views. The last group would say I had some good ideas but I have not fashioned a proper structure to carry them out. Americans can only move forward if the suggested policy can be embraced by a large majority over the long term (100 years or more). So, how do you fashion public policy that deals effectively with the problems of excessive GHG and does so in a world made up of such an extreme set of pluralistic world views and value systems?

The major thing we have going for us is that we are a representative democracy. We need to use its basic structure to fashion a stable approach to this extremely difficult problem. That is, to be responsive to the electorates.

These different belief systems are core to each individual and allow them to function in society. Each brings their world view to each situation, which allows them to interpret what they are seeing, what it means, and suggests the tools that they can use to cope with the situation. More recent work by cultural anthropologists indicates that there is little or no overlap of these belief systems, and to make public policy without acknowledging this as a basic reality is doomed to failure. The cultural anthropologists call this study area cultural theory, and they call fashioning an approach to try to include these disparate views, clumsy solutions [154].

The first time these concepts were applied to the energy field was in the early 1980s, and it proved a difficult frame of reference to base national policy upon. The original sponsor of the study [Wolf Haefele, the director of the energy programs at the International Institute of Systems Analysis (IIASA)] refused to publish the work and it finally was made public outside the institute [155]. At that time, Caputo coined the phrase "energy tribes" as a short hand to describe these different sets of attitudes, beliefs, and solutions. More recent work [155] is now acknowledged by the very same prestigious institute as a significant insight in obtaining viable political support for public policy and promoted it through its institutional magazine [157].

This cultural theory gives the promise of allowing a public policy to be fashioned that acknowledges the very different ways that people see "reality"; it gives a basis for policy given the very different worlds that we live in. Taking advantage of the quarter century for this policy framework to move from rejection to acceptance (at least in IIASA), these different political and personal world views can be summarized as follows.

There are four primary ways of organizing, perceiving, and justifying social relations; and they are named by the cultural theorist: egalitarianism, individualism, fatalism, and hierarchy. These four ways of perceiving life and its realities are in conflict in every conceivable domain of social life.

What is important to this study is that these views include the different ways in which people perceive and attempt to stave off a threat such as climate change. Let us examine this conceptual basis for understanding our political climate to see how we can use it to deal with climate change in a democracy.

In the egalitarian social setting, they see nature as fragile, intricately interconnected, and ephemeral. Man is seen as essentially caring (until corrupted by coercive institutions such as markets and hierarchies). We must all tread lightly on the earth. It is not enough that people start off equal; they must end up equal as well—equality of result. Trust and leveling go hand in hand, and institutions that distribute unequally are distrusted. Voluntary simplicity (conservation) is the only solution to our environment problems. People who feel this way usually call themselves core ecologists.

In the hierarchical social setting, people see the world as controllable. Nature is stable until pushed beyond discoverable limits; and man is malleable, deeply flawed but redeemable by firm, long-lasting, and trustworthy institutions. Fair distribution is by need, and the need is determined by expert and dispassionate authority. Environmental management requires certified experts to determine the precise locations of nature's limits and statutory regulation to ensure that all economic activity is kept with those limits.

In the individualistic social setting, people view nature as benign, resilient, and able to recover from any exploitation. Man is inherently self-seeking and atomistic. Trial and error in self-organizing ego-focused networks (unfettered markets) is the way to go. They feel that those that put in the most in should get the most out. Inequity is good and a natural part of the world of people (note the glib acceptance by entrepreneurials of massive benefits paid to top executives in the United States as just and right). They think that institutions that work with the grain of the market are what society needs.

The fatalistic social setting finds that neither rhyme nor reason makes sense in nature. Man is fickle and untrustworthy. Fairness is not to be found in this life. There is no possibility of effecting change for the better. Learning about nature is impossible. For them, a reasonable management response would be, "why bother" [157].

How is this view of the basic human social reality useful in any political situation? Well, if this is us, then we need to adjust any public policy framework to recognize this. Either we acknowledge this reality or we are doomed to failure caused by the approach that makes sense to only a subset of the public. For a public policy to be successful, it needs as a first step to acknowledge the way we are no matter how seemingly untenable, and then to go from there.

It appears that cultural theory has several normative implications [158]. People are arguing from different premises and will never agree. Each way of organizing and perceiving distils certain elements of experience and wisdom that are missed by the others. Each needs the others because each is incomplete and only represents a part of what is needed [159].

For example, under pure egalitarianism (core ecologists), there is an endless search for consensus. There is no official leadership that can settle issues or voting mechanism that can be evoked. This lack of procedures for settling conflicts or differences of opinion can easily paralyze the egalitarian social setting. In addition, pure egalitarianism creates social ill by ruling out any activities that would give rise to inequality of condition. This limits economic production to a bare minimum. Clearly, this value system would have to be blended with others for society to function at all.

Hierarchy has a whole armory of different solutions to internal conflicts. Individualism preaches the right of each individual to live according to his or her own needs and wants without group interference. Together, these two provide many ways to increase the resource base of a people. Fatalism is useful for egalitarian organizations as it continuously replenishes the moral outrage that keeps such organizations together [160].

Hierarchy also needs others. Without the distrust of central control and the insistence on transparency that are prevalent within both individualism and egalitarianism, hierarchy would be apt to be prey to the classic problems of bureaucracy: corruption, arbitrary use of power, tunnel vision, lack of innovativeness, and moral fragmentation [161]. Unfettered individualism undermines itself because it does not include the means to enforce property rights as well as contracts nor does it have the means to check accumulating inequalities or recognize environmental damage. They need egalitarian-minded organizations to notice and protest mounting inequalities and environmental insults. It also needs the regulatory capacities of hierarchy to enforce property rights and contracts, as well as to organize the continuous redistribution of resources to maintain social stability [157]. With this as background, let us take a look at climate change through the eyes of cultural theory and see what it tells us and see if it is a useful frame of reference for policy. Three climate change stories will be presented through the eyes of the egalitarians, the hierarchists, and the individualist.

The egalitarians see the fundamental cause of excessive carbon emissions as a direct result of the profligate consumption and production of the North, the industrialized countries mainly in the northern hemisphere. The core difficulty is the obsession with economic growth—the driving force of global capitalism. This has not only brought us to the brink of ecological disaster, it has also terribly distorted our understanding of both the natural and social world. Global commerce leads us to desire environmentally unsustainable products while our real human needs go unfulfilled. These needs are living in harmony with nature and each other. Finally, global capitalism distributes the spoils of global commerce highly inequitably. The egalitarian heroes are those who see through the chimera of progress and understand that the fate of humans is inextricably linked to the fate of planet Earth. To halt environmental degradation, we need to address the fundamental global inequities. Their solution is that unless a policy or action can be proven to be innocuous to the environment, it should not be carried out. The affluent North will have to fundamentally reform their political institutions and unsustainable lifestyles. Rather than professionalized bureaucracies and huge centralized administrations, we need to decentralize decision making down to the grass-roots level. Doing with less is the key strategy. Taking part in protests, lobbying, issuing research papers, and interfering with the juggernaut of progress all play in role in the solution to the climate change problem (http://www.earthfirstjournal.org/efj/primer/index.html, November 17, 2003).

The hierarchical view of climate change is quite different. It starts with a view of the limits to economic and population growth. The continued long-term use of oil, gas, and coal would eventually wreak havoc on the ecosystems on which humans depend. They do not believe the world is about to come to an end, and there is enough time to plan a gradual and incremental change toward energy options that do not emit GHGs. The underlying problem is the lack of global governance and planning that would rein in global markets and steer them in a direction to protect the global commons. The villains are those who are skeptical of the view that global intergovernmental treaties based on scientific planning and expert advice are what is really needed. The solution is for all governments to formally agree on the extent to which future emissions should be cut, which countries should do so, how and when. These governmental agreements should be imposed on the multitude of undiscerning consumers and producers within their borders. One should recognize that this is the logic behind the 1997 Kyoto Protocol.

The third story is that of the individualistic bent who view the recent public notice of climate change as much-ado-about-nothing. Just another attempt by naive eco-freaks who believe that the world can be made a better place by wishing it so, and by international bureaucrats looking to expand their own budgets and influence. They are skeptical of the diagnosis of climate change itself and they are convinced that even if it is correct, the consequences would not be catastrophic nor uniformly negative. For example, they point out that more CO_2 in the air would make things grow better. Climate change is not all that bad. This is not a unique environment catastrophe in the making. Rather, it is where we have always been—faced with uncertainties and challenges that if tackled boldly by a diversity of competing agents, can be transformed into opportunities from which all can benefit. They suggest a number of physical mechanisms that would undercut the scientific findings that support human-activated climate change. They see nature as wonderfully robust and bountiful. The answer is innovative business as usual [162]!

6.5   BASIS FOR STABLE POLICY

These three stories make sense to those embedded in each cultural system. They are viewed as incredulous by those in the "other" energy tribes. They give a plausible but conflicting view of climate change. None are wrong in the sense that they are implausible or incredible. Yet, none is completely right. They each have elements of what is needed. Trying to eliminate one or more of these "stories" would generate an incomplete and partially effective solution. Even more important is that each of these voices represents a part of the political process. Without representing each of these distinct voices in democratic states would lead to a loss of legitimacy. Although these are contradictory perspectives on policy, none can be implemented on its own. Only innovative combinations of bureaucratic measures, risky entrepreneurship, and technological progress, as well as frugality and international solidarity could be successful.

Using cultural theory to gain an understanding of the different views we bring to the table of any public policy issue would suggest that to make things work for the long term you need to keep all players in the game. Excluding one or more views will not be effective in contributing to a comprehensive solution or to garnering political legitimacy. To do this is awkward and even labeled clumsy by some. It involves a noisy, discordant, contradictory dialogue that in the end needs to be responsive to all disparate parts. Success in using this approach to a difficult public issue would be a combination of public policy and entrepreneurship; and citizens' activities have contributed to the improvement of a pressing, collective problem without making something else worse [163].

An example of a failed policy that excluded all but the individualistic views is the George W. Bush administration's approach to controlling the price of oil. This is a significant problem as Chapter 2 points out and it is called it the first "hitting the wall." I believe that the sustained high price of oil it is ultimately caused by the peaking of global oil production. Outside of some moments of political peeve, the Organization of the Petroleum Exporting Countries (OPEC) has no interest in having the price of oil to too high or too low. Too high and alternatives will make an inroad. Too low and their profits are reduced. OPEC's approach to oil pricing is like the story of the three bears. They attempt to manipulate it to be just right. However, the sustained high price of oil will be the market response to continued demand while global production levels and then starts the long side down.

Bush and his administration, who function primarily as individualists, would not see any limits to production as a problem. Rather, they viewed the high and variable price of oil as solely due to constraint of market forces by the OPEC cartel. They viewed the problem as OPEC's manipulation, and they viewed the solution as destroying OPEC's cartel. This would be done by finding a way to invade a suitable OPEC country, Iraq in this case, taking over the oil fields, significantly increasing preinvasion oil production with help of U.S. oil companies. This would flood the oil market and drop the price, thereby destabilizing OPEC. The more politically insecure members of OPEC (where excess oil profits were buying political stability) would be ripe for U.S. oil companies taking over their production when they could no longer pay the bills (see Chapter 2).

By not inviting other points of view of the problem and of solutions, Bush's policy was doomed to failure. By not opting for the "clumsy" solution, and by blocking other voices, he has stumbled badly wreaking enormous damage in so many ways. He might consider the resulting mess as collateral damage to his heroic attempt to restore the core individualist's credo—the free and unfretted market. The core problem was his belief that he along understood the problem and only he could fashion a clear and direct solution—the military option. Other points of view were unnecessary.

6.6 CAP AND TRADE

The European Union (EU) launched its initial attempt at the climate change cap and trade as part of the Kyoto Protocol in the Phase 1 agreement (http://ec.europa.eu/environment/climat/pdf/bali/eu_action.pdf). Initially, the Europeans were badgered into using a cap-and-trade approach by the Clinton administration who wanted a market-based approach to attract Republican support. The primary reason to attempt to structure an agreement that starts with the economically strongest nations and then extends to include additional nations and additional GHGs, is that limiting the agreement to a subset of nations would create a situation where the cost of energy and goods would be cheaper for those outside the agreement. This would lead to exporting industry and jobs to nations with carbon emission controls.

The EU approach was designed based on the Marrakech Accords of the Kyoto Protocol helped by the experience gained during the running of the voluntary UK Emission Trading Scheme in previous years (http://www.defra.gov.uk/Environment/climatechange/trading/uk/index.htm). Note was also taken of the U.S. cap-and-trade scheme on acid rain ingredients that has proven successful (http://www.edie.net/news_story.asp?id=6314).

The reluctance of largest GHG polluter, the United States since 1994, to engage in the negotiations has had a negative effect on the global expectations of success. At this point, almost none of the EU governments that have ratified the treaty have actually been fulfilling their requirements. Some critics point out that the Kyoto Protocol and the resulting EU Phase 1 agreement is based on the assumption that the prevention of climate change can only be provided through a formal, binding treaty between all governments. Furthermore, they say it has not identified and promoted competitive processes to help deal with climate change that can be much less costly or even a profitable undertaking [164]. They claim that the current Kyoto Protocol approach is based on only one way of valuing, the hierarchical one. This strictly technocratic and bureaucratic approach is at the root of many shortcomings of the Kyoto Protocol. These critics say that even if the Protocol is implemented, it will not prevent much global warming, stimulate economic growth, or empower destitute people. The cutbacks stipulated in Phase 1 are so small as to be insignificant compared to the worldwide 2050 goal of 50% reduction in GHG set by most scientists to stabilize the world climate.

Some of these criticisms have some weight. The Kyoto Protocol is cumbersome to implement and allows for three international implementation mechanisms: "international trading of emission permits", "joint implementation", and "clean development mechanism." The emission permit trading allows parties to comply with legal obligations to buy extra emission permits if it is cheaper than other ways they could reduce emissions themselves. The joint implementation scheme allows industrialized countries to reduce their emission through projects undertaken in other industrialized countries. The clean development mechanism allows industrialized countries to meet international obligations by helping developing countries reduce their emissions. All three international schemes are difficult to implement because they require extensive monitoring and complex calculations.

The trading permits reward countries that have poor economic performance and penalize those who are economically more successful, and the permit scheme is expensive to implement [165]. Because the permits are allocated to individual companies, you have to judge whether individual firms are in compliance. This requires monitoring and high transaction costs [166], and governments may choose to avoid this cost by letting the companies self-monitor. This will open the door to cheating and fraud.

In addition during Phase I, most emission allowances in all countries were given freely (grandfathering). This approach has been criticized as giving rise to windfall profits, being less efficient than auctioning, and providing too little incentive for innovative new competition to provide clean, renewable energy.

The joint implementation and clean development mechanisms requires the establishment of a baseline. This would predict what future GHG the company involved would have been if the company had not received foreign funding. Agreeing on a baseline is difficult [167]. There also is a financial incentive to overstate the amount of emission that will actually take place. Some form of oversight on the proper implementation of these two schemes must be implemented. Finally, it is an open question if the Kyoto Protocol could even be expanded and includes all the countries left out of the first set of agreements.

Most players in the negotiation see the Phase 1 agreement to be a first step—a small first step. Given the complexity of a system designed to monitor and reduce a core economic commodity such as carbon, it makes sense that the first step should be a small one. Setting up the structure even in organized countries such as the EU is quite a challenge. This will be the first time most of the nations on the planet would be engaging in a joint program to reduce and control such as major commodity.

After this initial step, the EU in January 2008 has decided on major revisions to overcome some of the initial shortcomings. These changes include centralized allocation (no more national allocation plans), auctioning a greater share (more than 60%) of permits rather than allocating freely giving historical polluters a massive windfall, and inclusion of the GHGs nitrous oxide and perfluorocarbons to the other four primary GHGs. Also, the proposed caps are to be increased and are aimed at an overall reduction of GHGs of 21% in 2020 compared to 2005 emissions. These address a number of the shortcomings of the first version of the Protocol, but some major issues remain. Is this cap-and-trade approach based solely on an approach that would be inside the world view of only one of the energy tribes—the hierarchical? The complexity of some of the mechanisms, and the high transaction cost for monitoring, opportunities for cheating and political manipulation, and finally, enforcing the agreement would lend support to this claim.

Cap and trade would also have a variable impact on consumers' power bills. During summer peaking loads in a hot spell or a really cold winter week, utilities would have to burn more coal to produce more power, causing their emissions to rise sharply. To offset the carbon, they would have to buy more credits, and the heavy demand would cause emission credit prices to skyrocket. The utilities would then pass those costs on to their customers, meaning that power bills might vary sharply from one month to the next.

That type of price volatility, which has been endemic to both the American and European cap-and-trade systems, does not just hurt consumers. It actually discourages innovation because, in times when power demand is low, power costs are low and there is little incentive to come up with cleaner technologies. Entrepreneurs and venture capitalists prefer stable prices so they can calculate whether they can make enough money by building a solar-powered mousetrap to make up for the cost of producing it.

Critics say a more effective, efficient, and equitable set of alternative policies may need to be developed. This would be based on involving all the "energy tribes" for all the reasons that the cultural theory suggests. This is especially important because a long-term approach needs to be developed because of the very long-term application of a consistent policy required to mitigate climate change. Only the involvement of all the different cultural value systems can maintain the politically stability needed for the long haul.

This criticism of cap and trade being hierarchical is curious because the major reason for a cap-and-trade mechanism is to set high-level standards that would achieve the needed reduction in carbon emission as determined by the planetary scientists and then allow market mechanisms to find the most cost-effective way to achieve this overall goal. This approach is supposedly aimed at using the market to make the thousands and millions of decisions each year to find the most cost-effective approach to reach a mutually agreed to goal. The shortcoming of the more traditional command and control approaches to environmental management is that it is often more expensive compared to the use of market-based economic incentives.

Cap and trade has been used successfully in the United States to address the acid rain problem in an economic manner. As illustrated in flawed Phase 1 of the Kyoto Protocol illustrates, the cap and trade must be well designed. Another example of a poorly designed program is the Bush Clean Skies program. It is not working because, "the cap is loose rather than firm, the governing rules are poorly designed rather than precise, the penalties for exceed the cap are low rather than significant, and the timetable for implementation is long rather than short," says Jeff Goodell [105]. When successful and failed examples of cap-and-trade programs are considered, what are the key elements that must be in the program? Using the outline provided by the Union of Concerned Scientists,

- Stringently capping emissions, with firm near-term goals. As discussed in Chapter 3, the United States must reduce its global warming emissions at about 80% below 2000 levels by 2050 to avoid the worst effects of global warming. Delay in taking action would require much sharper cuts later, making it much more difficult and costly to meet the necessary target. A near-term goal of about 20% reduction from 2005 levels by 2020 is essential.

- Including as many economic sectors as possible. The cap should cover all major sources of emissions, either directly or indirectly. They include electric utilities, transportation, and energy intensive industries, which together comprise some 80% of U.S. global warming pollution, as well as fossil fuel emissions from the agriculture, commercial, and residential sectors.

- Including all major heat-trapping gas emissions. Those include CO_2, methane, nitrous oxide, hydrofluorocarbons, perfluorocarbons, and sulfur hexafluoride. Emissions of different gases could be combined according to their global warming potential using the CO_2-equivalent method (http://www.epa.gov/climatechange/emissions/downloads06/07Annex6.pdf).

- Auctioning a substantial (60–80%) majority of emission allowances rather than giving them away to emitters. An allowance auction would allow the market to set the price of carbon, and it would be the most efficient and equitable way of distributing allowances. Giving away too many allowances would distort the market and could result in windfall profits for polluters.

- Using auction revenues for the public good. The government should invest auction revenues in clean, renewable energy technologies and energy efficiency measures. Revenues should also be used to ease the transition being stimulated in the following ways:

  o Compensate low-income families who have a larger part of their budget spent on energy
  o Provide transition assistance to workers or economic sectors that are disproportionately disrupted by the program, and
  o Help communities adapt to the unavoidable effects of global warming

- Excluding loopholes that undermine the integrity of the program. To be effective, a cap-and-trade program should not include a "safety valve" setting a maximum price for allowances and requiring the government to sell unlimited allowances to polluters once that price is hit. This would undermine the integrity of the emissions cap and reduce the incentive for investments in clean technology.

- Including strict criteria for cost-containment mechanisms such as offsets and borrowing. Offsets would allow regulated polluters to purchase emissions reductions from unregulated sectors or countries that do not have caps, instead of reducing an equivalent amount of their own emissions or buying allowances from other regulated facilities. For example, a regulated electricity generator could pay an unregulated landfill company to capture its methane emissions and use those emissions reductions to "offset" their own. Borrowing would allow facilities to emit more global warming pollution if they promise to make sharper emissions cuts later. Offsets and borrowing could lower the cap-and-trade program's short-term costs for polluters. However, by postponing emissions reductions from major emitting sectors, they would delay much-needed technological innovation and jeopardize the program's long-term goals. Any offsets should meet rigorous standards to ensure the activities are permanently removing carbon from the atmosphere beyond what would happen in a business-as-usual scenario. Borrowing should not reach unsustainable levels that threaten the program's viability.

- Linking with similar programs. There are important economic advantages to linking a domestic cap-and-trade regime with those in Europe and other regions that have adopted a stringent emissions cap. Doing so would require the U.S. program's design to be compatible with these other regimes [168].

Even if this list of elements of a good cap-and-trade program is established, it is clear that cap and trade needs a lot of monitoring, certification, and oversight. The timeline of the cap goals must be set (negotiated). The actual market sectors to be part of the system need to be determined. The GHGs have to be stipulated. The percentage of giveaway and auctioned emission allowances must be set. The distribution of auction revenue must be determined. No high limit on emission permit trading cost (safety valve) should be included. Strict criteria for offsets and borrowing imply a lot of verifying, oversight, and management. Linking to similar programs in other countries such as the EU would require a lot of negotiation and process management. The long-term administration and oversight of a cap-and-trade system would have to take place over a long series of national administrations far into the future. Some of these administrations, if similar to the Bush government, would be seriously at odds with the whole purpose of a cap-and-trade program and could use government department appointments to undercut the program. This technique was amply demonstrated by his effectiveness in undercutting the environmental program designed and built in the 1960s and 1970s in the United States. We really do not want an approach that would be too vulnerable to manipulation.

It is interesting to note the results of an attempt to integrate the views of industrialists and environmentalists from the U.S. Climate Action Partnership's attempt to form a unified approach to climate change. The main consensus points are as follows: the cap on the CO_2 should be applied as close to the point of emission as realistically possible; between 25% and 80% of the all emission permits should be given away to major emitter for a transitional period; the law should provide ample "offsets" available for purchase by companies failing to meet reduction targets; and "safety valves" should permit relaxed enforcement in case GHG reductions cause temporary economic hardship. These results were criticized because these are viewed as moving boldly in the wrong direction [169].

The point is made that to avoid the cap-and-trade program being overly complicated, that it should be applied where the carbon enters the economy rather than just before the carbon being emitted into the atmosphere. There are a much smaller number of carbon points of entry compared to the points of CO_2 release. The coal mine, the oil field, the ports, and the gas pipeline would be the 2000 points of entry. These entities would have to purchase emission credits from installations that have reduced their CO_2 emission. CO_2 emission reductions would take place at the auto factory and the coal power plant that increased efficiency, at the commercial building that was built to a Platinum LEED standard, etc. So you would still have many players that found it advantageous to reduce carbon emissions to earn credits, but a relatively few players at the source of the carbon entry into society would be purchasing these emission credits. It would simplify the bookkeeping.

The range of 25–80% for auctioning emission permits is too large. A substantial majority of these permits should be auctioned to avoid a windfall profit to carbon emitters. This range would have to be limited to 60–80% to limit the windfall profit. Even this 20% range of uncertainty would open the door to backroom dealing that would undercut the program. One of the major positive attributes of a cap-and-trade approach is that a cap is set on a timeline to get you when you need to go. The other is that the full complexity and innovation of the global economic system can be turned loose to find the enormous number of approaches that can be assembled to reach this goal in a economically effective manner (minus administrative costs and taken opportunities for fraud and cheating).

A revenue-neutral carbon dumping fee approach would have some of these features such as setting the tax and the reciprocal ways the tax would be redistributed to reduce the payroll tax and investment taxes. The GHGs that are part of the tax program would have to be identified. But once it is put in place, it would be vastly simpler to administrate and quite transparent. Manipulation would be difficult because there are a limited number of aspects of a revenue-neutral carbon tax approach and a tax is fairly transparent. However, to keep different nations on an even playing field, the magnitude of the fee would have to be similar for all nations. This assumes that an international negotiation takes place to establish the fee to be used by all. Although the structure of a single fee is very much simpler than the number and type of agreements in a cap-and-trade scheme, a fee requires a large number of nations to agree.

It is interesting that Gore recommended to Congress "that we should work toward de facto compliance with Kyoto. If we can ratify it, fine. But, again, I understand the difficulty. But we should work toward de facto compliance. And we have to find a creative way to build more confidence that China and India and the developing nations will be a party to that treaty sooner rather than later" [170].

At the federal level, the acid rain program provides a good model of a successful cap-and-trade program that has greatly reduced power plant emissions of sulfur dioxide (SO_2) and nitrogen oxides (NO_x), the pollutants that cause acid rain and smog. Besides the shaky start of the EU-implemented cap-and-trade program in 2005, there are two other closer to home cap-and-trade programs that are coming on-line. The Regional Greenhouse Gas Initiative (RGGI), which will begin in January 2009, is a cap-and-trade program designed to reduce emissions from the electric power sector in 10 northeast and Mid-Atlantic states (www.rggi.org). In addition, six Western states and two Canadian provinces have launched the Western Climate Initiative to develop a regional cap-and-trade regime, and several Midwestern states are proposing similar programs as part of climate change legislation (www.westernclimateinitiative.org). California, the 10th largest polluter in the world, has set a cap of reaching 1990 levels of global warming pollution by 2020, and is moving to implement a suite of policies, including an emissions trading system, to achieve that goal (www.climatechange.ca.gov).

Key to cap and trade is to have bureaucratic measures aimed at overarching goals but designed to unleash competitive and creative market processes to achieve these goals. These goals would have to be structured differently for industrialized, emerging developing countries such as China, India, and Brazil, and countries that continue to stagnate economically. The industrialized countries would have a goal like the United States to reduce GHG by 80% by 2050, whereas emerging countries would have to limit GHG growth rather than reducing it initially. There eventually would have to be a reduction goal set even for the emerging countries so that the world average carbon emission reduction is about 50% by 2050. The economically stagnating nations would not have active goals but would have opportunities to end certain carbon emission practices such as slash-and-burn farming by being part of offset projects.

In addition to cap and trade driving down GHG emissions, poor (South) counties would be given the opportunity to locally produce and consume cheap forms of distributed renewable energy especially when you consider avoiding the enormous cost of the conventional energy infrastructure. There are many opportunities for a North–South exchange including renewable technology and incentives to end carbon emission practices like cutting down equatorial forests.

6.7   REVENUE-NEUTRAL CARBON DUMPING FEE

Although cap and trade creates opportunities for cheating, leads to fluctuations in energy prices, and has high administrative cost, carbon fees can be structured to sidestep all those problems while providing a more reliable market incentive to produce clean-energy technology.

A carbon fee simply imposes a fee (tax) for polluting based on the amount emitted, thus encouraging polluters to clean up and entrepreneurs to come up with alternatives. The fee is constant or to be increased in known amounts over time and thus predictable. It does not require the creation of a new energy trading market, and it can be collected by existing state and federal agencies. It is straightforward and much harder to manipulate by special interests than the politicized process of allocating carbon credits. It does not have the difficulties in figuring out baselines, offsets, and borrowing. It does not have to verify the carbon emission reduction in various energy sectors in the United States or abroad.

And it could be structured to be far less harmful to power consumers. Although all the costs and benefits of the traded emission permits under cap and trade go to companies, utilities, and traders, the government does receive revenue from the auctioning of emission permits. Essentially, all the added costs (fees) under a carbon tax would go to the government over the years. This revenue source could be used for several things such as to offset payroll and corporate taxes. So although consumers would pay more for energy, they might pay less income tax or some other tax. The idea is that taxes are not going up—they are being restructured. Less taxes on things we like (payroll, profits) and more taxes on things we do not like (carbon emissions). We already are quite good at collecting taxes, and the institutional mechanisms are in place. There is no need for a new elaborate bureaucratic institution that must tackle difficult problems identified earlier. Although cheating is possible in any tax scheme, the opportunities are vastly constrained compared to a cap-and-trade program. Although taxes are always a political liability, the cap-and-trade approach will also incur costs that will be passed on to the public which is essentially a tax. With a revenue-neutral carbon emission fee, the increased cost associated with carbon emission activities will be balanced by the equal reduction in payroll taxes or a per person rebate.

There is a growing consensus among economists around the world that a carbon tax is the best way to combat global warming, and there are prominent backers across the political spectrum, from N. Gregory Mankiw, former chairman of the Bush administration's Council on Economic Advisors, and former Federal Reserve Chairman Alan Greenspan to former Vice President Al Gore and Sierra Club head Carl Pope. Yet the political consensus is going in a very different direction. European leaders are pushing hard for the United States and other countries to join their flawed but improved carbon-trading scheme, and there are no fewer than seven bills before Congress that would impose a federal cap-and-trade system. On the other side, there is just one lonely bill in the House, from Rep. Pete Stark (D-Fremont) to impose a carbon tax, and it is not expected to go far.

The obvious reason is that, for voters, taxes are radioactive, while carbon trading sounds like something that just affects utilities and big corporations. The many green politicians stumping for cap and trade seldom point out that such a system would result in higher and less predictable power bills. Ironically, although a carbon tax could cost voters less, cap and trade is being sold as the more consumer-friendly approach. A feature of a fee approach is that you are not actually reducing the amount of GHGs on a certain timeline as in a cap-and-trade to reach the goal that scientists determine. You are directly increasing the cost of emitting GHG and as a result you expect a reduction in emissions as energy efficiency, switching to alternatives and a whole hose of other things takes place. You would have to monitor the actual emission reductions and adjust the amount of the fee to achieve the results desired.

A well-designed, well-monitored carbon-trading scheme could deeply reduce GHGs with less economic damage than pure regulation. It would be so complex that it would probably take many years to iron out all the wrinkles. Voters might well embrace revenue neutral carbon emission fees if political leaders were more honest about the comparative costs [171]. The revenue-neutral carbon tax could be applied to all fossil fuel sectors including vehicle fuels even if a strong Corporate Average Fuel Economy (CAFÉ) standard is put in place. An aggressive CAFÉ standard that is incremented periodically would achieve the desired reduction in carbon emissions in the vehicle based on increasing efficiency. However, the second problem with vehicle use is the increasing mileage driven each year. Since 1983, the United States has added more than 60 billion miles per year to road driving, which is about 2% per year (Federal Highway Administration Office of Highway Policy Information Monthly). Even if increased CAFÉ standards improve vehicle efficiency, these gains will be eroded by the increase in miles drive. There needs to be an addition element beyond CAFÉ standards such as a revenue-neutral carbon dumping fee applied to petroleum products.

Beyond the CAFÉ standards, there is a strong need to give attention to a range of other issues such as improving public transport and reversing our approach to land development to reverse sprawl. Both of these approaches are important but will take decades to slowly put in place.

The revenue-neutral carbon dumping fee looks like it does a better job when viewed through the eyes of cultural theory. It does not create such a bureaucratic overlayer in an attempt to maintain the trading system and avoid gaming the system with various types of manipulation. Entrepreneurs should be giving encouragement to improving the process and design of innovation as solutions, not how to manipulate the system. The cost is predictable and business can operate with more assurance with a fee. They will know the rules and the rules will be less amenable to political manipulation once established. It is the policy for the long haul in a multipluralistic world. The cap and trade is less so.

6.8   PARALLEL POLICIES

Whichever approach is used, additional policy is needed to further level the playing field for renewables and to more quickly capture the increasing economics of renewables. For this global transition to take place, the basic economics must be favorable and competitive for it to happen at all [59].

As this study shows, the basic economics are favorable. The main barriers to rapid introduction of cost-effective energy efficiency and renewables in the United States are the current investment in conventional energy; the powerful political influence of oil, coal, gas, and nuclear interests; and the structure of the energy system that is based on conventional energy.

A revenue-neutral carbon tax or a cap-and-trade program alone would not be sufficient to meet the challenge of climate change. Although both would address the failure of the market to account for carbon emissions that harm to the climate, it cannot by itself provide sufficient incentives for the technologies and other measures that will be needed to establish a true low-carbon economy. Parallel policies are needed to ensure development and deployment of the full range of clean technologies. These policies include requiring utilities to generate a higher percentage of their electricity from renewable energy sources such as 20% by 2020 (Renewable Energy Portfolio Standards); requiring automakers to increase vehicle fuel economy standards (CAFÉ) such as 40 mpg by 2015 and increase by 3% per year after that; stronger energy efficiency policies such as a national building code modeled on California's Title 24 with carbon-neutral new construction by 2025; incentives for investments in low-carbon technologies (investment and production tax credits); and policies encouraging smart growth [168].

Studies have shown that a comprehensive approach including these parallel policies would lower the price for GHG emission allowances, cut emissions, and save consumers money by lowering their electric and gasoline bills. Office of Management of Budget (OMB) examining the McCain–Lieberman bipartisan cap-and-trade legislation (S139) for example and found that the economic results of this legislation tended to stabilize fossil fuel prices and accrue economic benefits to citizens. This amounted to $48 billion per year savings by 2020 and is in addition to the $82 billion saved by the energy efficiency and renewable energy strategy developed in Chapter 4. This is a total saving of about $124 billion/yr using this noncarbon strategy. The parallel policies included were renewable transportation fuels standards, renewable portfolio electricity standards, incentives and barrier removal for CHP systems, caps on other power plant pollutants—SO_2, NO_x, and mercury (as in S.843), and smart growth measures. Apparently, a functioning cap-and-trade program with reasonable parallel policy measures can reduce energy costs [172].

6.8.1   Research Investment

A strategy to move toward the reduction in carbon emission would entail a number of measures. To capture the fuller technical/economic potential of renewables, it is vital to redirect and significantly increase the RD&D budgets of the dozen or so countries that current account for 95% of the world energy RD&D [173]. Fortunately, the United States has enormous capacity to invest in RD&D and can be matched collectively by the rest of the industrialized countries. In 1999, the U.S. federal government alone directed $4.4 billion at all energy RD&D. At the same time, it poured abut $40 billion into military RD&D. Given that U.S. spending on the military (not counting the Iraq and Afghanistan wars) is about the same as all the other 190 countries in the world combined and the enormous U.S. military superiority, there is a significant opportunity to apply some of this misdirected military spending into the energy area. One can easily envision increasing the total energy RD&D by factors without affecting the U.S. overwhelming military superiority. Military superiority is not proving especially useful in the current Middle Eastern military engagement. If anything, a very superior military tempts certain politicians to use it inappropriately. In addition, by redirecting the DOE energy RD&D funding from conventional to renewable energy, it would make a real difference in bringing needed future advances into being quicker.

This also implies the removal of supports for conventional (oil, nuclear, coal, and gas) energy, which will not get us where we need to go and transfer this funding to renewables. This is easier said than done because of the vested interests of conventional energy giants. The key areas where RD&D would make a major difference in the renewable energy areas are PV for increased performance and lower cost, geothermal to develop Enhanced Geothermal Systems and technology to overcome limitations of depth, relatively low permeability, or lack of water as a carrier fluid for the heat energy, to renew the Production Tax Credit of 1.8 cents/kW·h for wind systems until 2010 and then smoothly phased out by 2020. A 30% investment tax credit for concentrating solar power system (CSP) to 2017 and then smoothly phased out by 2025. Biomass-electric systems are mature and only needs some stimulus to set up the small power plants in areas with ample agricultural, municipal and forest wastes with an adequate collection system. This support could be in the form of an investment tax credit as in the CSP systems. There is a great need to stimulate RD&D in the biofuels arena that is concentrated on lignocellulosic biomass as the feedstock. When the R&D is done, then development is needed to establish commercial feasibility. Stimulus could be provided by re-directing of the ill-advised corn-to-ethanol program price supports.

The projected gains in the building energy efficiency sector depended to some extent on RD&D in areas such as solid-state lighting, advanced geothermal heat pumps, integrated equipment, efficient operations, and smart roofs.

One renewable option not consider but one that deserves RD&D support is ocean energy. This is a large resource and located off-shore near many urban centers. Development is needed to achieve cost effective systems that can withstand the harsh ocean environment.

However, there are several examples where some RD&D for conventional energy sources is appropriate such as identified in earlier chapters for coal and nuclear and the current RD&D could be redirected to be more useful. For example, coal needs certified long-term sequestration sites if coal is to play a role in reducing carbon emissions, along with the demonstration of a number of new coal systems using different types of coal to provide the impetuous for the coal industry to make the transition to "clean" coal as discussed in Section 5.1.

Nuclear has a range of needed changes such as a new long-term waste disposal site or sites, removal of on-site wastes now in water pools to protected dry canisters in the interim, revamping of the NRC to restore its safety culture, actually insisting on new reactors being safer, that terrorism threats be explicitly considered by the NRC, that fast reactors and reprocessing nuclear wastes into weapon grade materials, along with a number of other specific changes listed in Section 5.2.7. The lion share of RD&D logically needs to be focused on all the renewable technologies to hasten the longer-term economic efficiency of these energy systems. Chapter 4 identified many specific RD&D issues with longer-term renewable technology opportunities.

It is important that the RD&D allocation be distributed among all technologies with potential to help the United States move away from carbon. For example, it was a mistake in the post-World War II era to invest almost exclusively in nuclear power. Finally, it is vital not to channel all public funds through one agency. Institutionally dynamics often lead to a point where blind spots emerge and distortions occur that are not self-correcting. Multiple RD&D agencies will ensure that institutional myopia is minimized [168].

Along with this increasing and refocusing of RD&D efforts, there is a government role to mobilize capital, to adapt infrastructure especially long-distance electric transmission, to shift taxes and subsidies, to provide resources to train installers and maintenance workers, encourage local governments and states to actively remove barriers to renewable energy systems and to engage in a dialogue with companies and citizens' groups to grasp the opportunities of energy efficiency, renewables, and conservation [157]. Some renewable options have protracted conflicts usually over siting issues. As in the case of the Nantucket Sound wind farm conflict, it proved useful to engage a public participation process that brought all parties together and allowed interaction with the project. The results proved satisfactory to about 80% of the participants [174]. This type of process is recommended for all renewable energy systems with likely public conflicts.

As with new coal plants, there would be a government role to stimulate new types of renewable plants to accelerate their acceptance. This is especially true for geothermal, some CSP plants, and some biomass systems. Early new types of plant could receive special incentives to assist the transition from advanced prototype to full-scale commercial systems in various region specific situations.

6.8.2 National Electric Transmission Grid

Some forms of renewables such as wind, concentration solar, and geothermal have a large national level resource, but only some of this potential is in near large urban load centers. These near urban center opportunities will be well on the way to saturation by 2030. Around 2025, initial links in a national electric transmission system are needed that will allow these large renewable resources to start achieving their national potential. Infrastructure upgrades are needed that are beyond the normal and need to be addressed by the national government. After initial planning, this effort needs to get underway by 2015 to meet the goal of building the initial long distance links by 2025.

The national investment in a long-distance high-voltage DC electric transmission system is needed much as the Interstate Highway System was seen as a wise investment in the greater economic integration of the country. This would allow the efficient movement of CSP electricity from the southwest (seven states from western Texas to southern California) to distant urban centers. In a similar manner, wind from 20 states of the Midwest and West plus some coastal states, geothermal from western (16 states primarily from Louisiana to Washington), and agricultural states and the Southeast biomass wastes to electricity systems could be used in distant cities.

These four energy systems compliment each other beautifully in that they produce electricity around the clock. Geothermal and biomass-electricity are baseload (24/7), whereas some types of CSP (parabolic trough and central receiver) are a day-time electricity generator that can have inexpensive thermal storage on-site and be a midrange energy producer and operate comfortably up to 12 hours per day at rated power. Some CSPs (concentration PV and dish-Stirling) would generate electricity only during the bright sunlight hours (up to an average of 8 hours per day) and generate more daytime peaking power. PV on buildings is a midday energy generator. Finally, wind is more of a reciprocal to CSP generator and would augment the nighttime loads especially the emerging new large transportation load of pluggable hybrid-electric vehicles (PHEVs). These vehicles will initially allow the commuting mileage to be transferred to nighttime electricity in a much more cost-effective manner than using food (corn) as a source of vehicle energy. For the western renewables, the time zone difference of up to 3 hours would act like 3 hours of storage to extend power into the early evening.

This remarkable renewable energy combination is extraordinary and especially so if these large resources are connected via an efficient electrical grid that can be created only with significant support at the national level. The railroads were built with a strong role of the federal government, although it was a thoroughly entrepreneurial undertaking. A similar approach is needed in the creating of a truly national electric energy system.

In addition, there are some key reforms needs in the transmission area. An example is to institute a new innovative transmission tariff to provided long-term transmission access on a conditional firm or nonfirm basis. Such a transmission tariffs would speed the development of wind and other renewables and increase the efficiency of the transmission system.

Develop smart grid systems across the nation for a number of reasons but also to facilitate the nighttime charging of PHEVs as an interruptible load to allow greater use of wind power in a region.

Expand FERC Order 888 to explicitly ask individual transmission operators to offer alternative nonfirm service for periods longer than 1 year. Nonfirm service is not guaranteed, so service can be interrupted under specific curtailment procedures and priorities and would allow wind. Alternative tariff could be conditional firm. The main characteristic of the conditional firm tariff involved a cap on the number of hours that the generator would be curtailed, or a long-term nonfirm tariff. This would give some renewable technologies greater access to the existing grid before the national grid is developed.

On-site renewable energy systems including PV, smaller version of CSP systems, hot water, solar air conditioning, on-site solar steam, and urban biomass-electricity systems generating electricity inside the urban center would not be part of the national electric transport system. As shown earlier, these all pay a significant role in growing our energy future.

6.8.3 Energy Efficiency Market Failure

In the vehicle and other energy efficiency areas that have such enormous potential, there is definitely a need for a policy role. There is a systemic undervaluing of life-cycle costs in the designing of vehicles, buildings, and industrial processes. To capture the creative entrepreneurial skills and yet achieve the cost-effective adjustments needed, industry-wide standards can and should be used. CAFÉ standards for vehicles were discussed earlier as a simple and effective way to increase efficiency in vehicle design that is cost effective over the life of the vehicle. Current technology can easily achieve 40 mpg by 2015. Improvements in current technology and improvements in pluggable hybrid-electric and eventually hydrogen-driven fuel cells could continue to deliver gains if driven by periodic CAFÉ standards upgrades (3% per year). A similar story can be made for aircraft since today fleet was not designed for $150/bbl or greater fuel costs. An interesting fuel for advanced commercial aircraft is liquid hydrogen, which could be independent of fossil fuels [175].

To support vehicle efficiency in addition to the evolving CAFÉ standards, a revenue neutral "feebate" program can be instituted where cars that achieve greater than then average CAFÉ standard are given a rebate of about $500 for each mpg over the standard. This rebate would be paid for by a $500 fee for each mpg under the standard at the time of the vehicle sale [176]. Alternative fuels should be supported by RD&D to bring superior techniques to market and with initial subsidies for new plants and facilities. However, the current practice of allowing dual-fueled vehicles mpg to be counted at double their actual gasoline based mpg rating is counterproductive. As currently implemented, the manufacturers are installing dual-fueled capability on vehicles with some of the poorest mileage characteristics. The dual-fuel capability is installed not because it is expected that these vehicles will use the biofuels; it is only being done to be able to build more of these gas guzzlers than would normally be allowed under CAFÉ standards. Other approaches exist to encourage dual-fuel capability without supporting this sham.

Public funding for public transport needs to be increased significantly, along with support for land use that helps public transport work more effectively. Smart growth has been talked about with great enthusiasm for decades as an alternative to suburban sprawl. The results to date are discouraging. It is time to take a serious look at what has been tried in different cities and pick a few winners, and then to design an approach that sets up policy that the private sector can function within and get where we want to go.

The equivalent of CAFÉ standards can and should be developed for buildings (residential, commercial and industrial). An example at the state level is Title 24 in California building code that has been used for the last 30 years to reduce building energy use throughout the state. California now used half the electricity per capita as the average of the nation that in large measure is due to the extending the building code to housing energy use. Title 24 requires that a particular building type use less than a certain amount of energy per square foot per year based on certified engineering estimates. This certification is needed for a building permit to be issued, and this idea can be taken to a national level. Key to this building energy efficiency standard is that there are no specifications on how this energy requirement is met. It is up to the architect/builder/owner to choose what collection of techniques works best and is most cost effective in their region and meets their esthetic tastes. The Vermont zero carbon program also has important lesions that can be taken national.

An interesting recommendation from Al Gore in the housing area is that we ought to set up a carbon-neutral mortgage association where all of the extra carbon reduction costs in new construction are set aside. They will pay for themselves in lower energy bills. But just like Fanny Mae and Freddie Mac, put them in an instrument that is separate from the base purchase price of the house. When you are closing on a house and you sign the mortgage, and they will say here is your Connie Mae home improvement package. You do not have to worry about paying for that because it will pay for itself [170].

This sounds close to a suggestion by the city of Berkeley to pay for home upgrades on existing homes that reduce energy use. In Berkeley's case, the city will provide the capital to finance the cost of the energy improvements with low cost municipal borrowing and this will be paid back via increase taxes on the property for a fixed period. Then, the home owner receives the benefit of reduced energy expensive without having to raise the capital.

This combination of buildings and vehicle policies that leave the entrepreneurials free to develop the techniques that are sensitive to the local markets and conditions seems to be an effective approach based on cultural theory. That is, to have the needed national policy developed and turning loose the individualistic value system to figure out how to actually achieve it. This approach that is not dominated by only one energy tribe and would be more effective in the long run than solutions that depend on only one way of thinking.

Another key element of California maintaining its electricity use per person at a constant value over the last 30 years in spite of increase home size greater use of electrical devices, was the use of appliance efficiency standards. It would be wise to increase federal funding and periodically update appliance efficiency codes and standards and expand the Energy Star program.

Compact fluorescents have developed sufficiently with several colors of light (warm, blue, and in between), different shapes and sizes, and indoor and outdoor versions at prices that are sometimes under $1 per bulb in a bulk package. It is time to phase out the incandescent light bulb completely. This transition to compact fluorescent lightbulbs and solid-state LEDs will increase the energy efficiency of a particular level of light by a factor of 4 to 10. Part of this would be requiring that lamp manufactures make adjustments to accommodate typical slightly larger compact fluorescents in order to sell in the US markets. This recommendation is one of the few that relies on just the hierarchical value system, which is a specific technical solution imposed on the market. Although normally to be avoided, an exception is made in this case. As part of meeting the California building code Title 24 requirements, buildings do have some designated hard-wired fluorescent lights. Compact fluorescents are not included as part of a building code because they can be interchanged with incandescent bulbs by the building occupant. It is time to phase out incandescent over the next 5 years.

The third leg of the California success in not increasing its per capita electricity use for 30 years was the use of a wide range of energy efficiency programs using the local utility to run the program using public monies. There has been a mixed bag for these varied programs and often the utility is less than enthusiastic about saving energy. By decoupling the utility incentives so that they are rewarded for actively supporting energy efficiency measures, it would be possible to encourage the utilities to run programs more successfully. It also makes sense to open up these programs beyond the for-profit utilities and the municipal utilities. Bidding for these state contracts should be opened to nonprofit (such as the California Center for Sustainable Energy) and nonutility for-profit companies to more effectively use these public monies.

Time of use electricity pricing needs to be instituted throughout the country to give the consumers a clear market signal. They will be paying what the electricity actually costs to deliver for each hour of the day. In conjunction with this, smart meters are needed for a host of reasons. One vitally important reason is to be able to mound a display inside the home that shows the current energy use, current price, as well as a selection of other metrics to give immediate feedback to the building user about current use and cost. Other smart meter features is to have certain appliances on interruptible service at a more favorable price to allow active load management by the utility. A smart metering system can also accommodate remote electrical outlets that can permit electric vehicle charging to a particular customer using a credit card swipe for identification and billing. Also, PHEVs can plug in throughout the grid and provide grid backup in the form of both standby power (for reliability purposes) as well as to actually pump energy into the grid. Again, the swipe identification card will tell the utility who to send the check to at the end of the month for the energy services provided [177].

6.8.4   Energy Structure Blocks Renewables

Energy prices are heavily dictated by existing infrastructure for generating, distributing, and consuming energy. They are also dependent on many public institutions such as state and federal government regulation of energy markets. As a result, the government has quite a bit of influence over energy prices. Currently, these public institutions are structured toward conventional energy systems.

An example is the Public Utility Commission in California (CPUC), which regulates the private utilities such as the San Diego Gas and Electric (SDG&E) Company. It is interesting that installing an on-site PV system on a commercial or school building in San Diego that generates most of its own electricity will actually raise the electricity bill for most of these buildings. This certainly seems counterintuitive. It has to do with the CPUC allowed rate structure that SDG&E uses. A combination of fix charges, capacity charges, demand charges, and so on, drive the cost of electricity in some applications to be more with the PV system generating most of the on-site electricity than without the PV system at all. (After a lot of public criticism triggered by the electric bill going up for schools after installing a large PV system, these rates have recently been adjusted to be more favorable for an on-site PV system.)

Similar difficulties exist in many utilities for clients that install a CHP system on site. This type of system (CHP) that uses the waste heat from the electricity generation for on-site heating applications can double the efficiency of the fuel used. CHPs more than halve the carbon emission if fossil fuels are used, and it also offers economic advantages compared to buying both electricity and fuel for just heat. One would think this is a desirable system where even the specific site conditions are favorable. However, the rate structure of many utilities around the county penalizes this very attractive approach. So there is a significant role for governmental public utility commissions to examine the current regulatory structure and clean house. It is necessary for all state regulatory agencies as well as the federal agencies to make all the adjustments that encourage reduced carbon emission systems and, if anything, penalize high-carbon emission systems in their jurisdiction.

There is a role for the federal government that would hasten investments in carbon-neutral activities. The Securities Exchange Commission ought to require disclosure of carbon emission in corporate reporting. This would more quickly give information to investors about which corporations are more or less vulnerable to future carbon costs [171].

What are other policy mechanisms that make sense based on the insights of the cultural theory? There are a number such as a national renewable energy portfolio standard. An example would be a national goal of 20% renewables by the year 2020. States could have a more ambitious standard if they wish such as the 33% goal in California. This standard would apply to all private and municipal utilities as well as rural coops. Again, how to meet this standard is left up the utilities teaming up with individualistic energy providers. This goal can be raised periodically in a prudent way to steer the economy toward the U.S. 80% carbon reduction goal by 2050.

To expedite this goal, a number of supports are appropriate in the near term. Examples are either an investment tax credit or a production tax credit for all renewables. The magnitude of these tax incentives should be reduced over the next 20 years. This is a two-decade policy to help bridge the transition to renewable energy. The start/stop production tax credit that has characterized the past and current energy policy is really counterproductive. This type of uncertainty in the basic supports for renewables is incredibly disruptive to the energy market.

Upfront capital rebates are also a mechanism to encourage early energy systems, and it can be phases out gradually as market costs are reduced by increased production and learning curve efficiencies (see http://www.cpuc.ca.gov/puc/Energy/solar/). An alternate strategy is to pay an energy feed-in tariff for each unit of energy generated by renewable energy. Again, this can be used to stimulate renewable energy in early markets and phased out as commercial system decrease in cost.

Establishing a set of solar rights nationally to prohibit the infringement of building owners to access their solar resource could be based on successful state examples such as California. Also, having solar rights in place that are not infringed by local covenants such as municipalities or home-owner associations is very important [178].

When an on-site solar electric system is installed that is connected to the grid, this connection is called net metering. Excess electricity is pumped into the grid for sale by the utility to other users (usually during the day), and it is provided by the utility when electricity is needed (usually at night). The utility usually gets electricity during high prices times of the day and provides electricity for the most part during less expensive off-peak times. The utility usually can defer or avoid electrical distribution system upgrades because there are small-generation plants distributed along the lines. The home user buys and sells electricity at the same retail rate and does not have to put in an energy storage system. This is a win-win. It is important to establish national net metering and interconnection standards for on-site renewable energy technologies. The current mish-mash across states and counties makes it very difficult for renewable energy providers to function. Also, the net-metering law should not be limited to the energy used on-site. If extra roof is available beyond the needs of the building, then it should be up to the building owner to decide whether more of the roof is used to generate power. The amount that the building owner is paid for the excess electricity should reflect the time of day that it is generated. Allowing the full roof to be used will essentially double the on-site PV resource potential.

6.8.5 Conservation: Green Spirituality or Common Sense

Energy conservation did not play a role in the numerical projections of this study. Conservation is the vast array of voluntary actions that citizen and corporations can take to reduce their energy consumption. These can be accomplished during a relatively short time during an energy crisis or in the long term as a lifestyle change. A range of motivations could spur conservation from a civic response to an energy emergency, to cost savings, to increasing environmental consciousness, to a core personal belief that makes a person one with the universe (egalitarians).

This important energy option is a key strategy of the egalitarians, and as a matter of government policy, it should be encouraged at every opportunity. Because a substantial number of citizens see conservation as vital, programs to bring them together physically and electronically should be developed. Workshops, conferences, seminars, books, magazine articles, blogs, and so on, should be encouraged as part of a government-supported programs. This needs to done recognizing that this makes enormous sense to some Americans and not others. The program design needs to recognize and try to avoid presenting these ideas as something for everybody and make sure the design of the program is aimed at those who would understand and use these ideas. Others in society with a lesser intensity of their motivation for conservation could also be encouraged as a part of our overall approach to sound energy policy. The potential benefit in emission reduction is potentially enormous and difficult to estimate without some program results.

The climate change crisis is at its very bottom, a crisis of lifestyles. Some might even say it is a crisis of character. The United States is the third largest oil producer on the planet (after Saudi Arabia and Russia). Yet we do not live within this very large production capacity. A similar thing can be said about the size of our houses, the size of our cars, and so on. Why is it not reasonable for us to live within our energy means? Why is our culture so dependent of spending as an act of virtue? There could be an appreciation of a simpler life with other virtues besides spending money. The big problem of overloading the planet's atmosphere is nothing more or less than the sum total of countless little everyday choices. Most of these are made by us (consuming spending represents 70% of our economy), and most of the rest is made in the name of our needs, desires, and preferences. What would happen if each of us went green?

Sometimes you have to act as if acting will make difference, even when you cannot prove that it will. This, after all, was precisely what happened in Communist Czechoslovakia and Poland, when a handful of individuals such as Vaclav Havel and Adam Michnik resolved that they would simply conduct their lives "as if" they lived in a free society. That improbable bet created a tiny space of liberty that, in time, expanded to take in, and to help take down, the whole of the Eastern bloc [179]. Conservation warrants serious attention as an energy option. It would take a combination of very creative government-supported policy and a grass-roots campaign to ignite a broad back-to-green movement that could sweep this county and the world. Why not?

6.9   WHAT TO DO WITH COAL PLANTS

The 800-lb gorilla in this chapter that has not been directly faced is the question of what to do about coal plant carbon emission? Will a revenue-neutral carbon tax or a cap-and-trade program handle it? Section 5.1 logically looked at the coal situation and came up with a coal strategy that:

- Increases RD&D for CO2 sequestration.
- Establishes a national Carbon Sequestration Commission to certify sites for long-term CO_2 impounding and give oversight to the national sequestration program.
- Suggests government support for the demonstration of the newer technologies (IGCC and oxy-fuel) at full scale with CO_2 sequestration in different utility and geologic environments with different grades of coal.
- Bans all new conventional coal plants without sequestration at certified sites.
- In the near term, encourages efficient natural gas combined cycle (CC) plants to replace older coal plants and to be used instead of new traditional coal plants. (This type of gas plant generates less than half the CO_2 of today's coal plants at a cost less than a new coal plant with sequestration.)
- In the near term, encourages biomass co-firing with coal.
- Encourages efficient existing coal plants that happens to be sited in a good CO_2 sequestration area to be retrofitted with the oxy-fuel process that allows clean operation and the separation of CO_2.
- Encourages new IGCC and oxy-fuel coal plants to be built near good sequestration sites to finally retire all the remaining old coal plants.

This certainly would do the job, but the problem with this is that is uses the words "ban" and "encourages." The question is how? How to do design an approach that uses cultural theory and present a strategy that includes the various views of reality? The first three elements of the strategy are all federal government roles to invest in sequestration RD&D, supporting new large plant demonstrations, and setting up a new commission to oversee the certification of sequestration sites and their long-term stability. However, the final five elements are all top-down command and control solutions. This is the type of thing that the hierarchy social setting would support without any involvement of the individualistic or the egalitarian approach and would be a poor policy approach.

The cap-and-trade policy mechanisms suggested earlier would certainly put a cost on carbon emissions that would "encourage" the substitution of combined cycle gas plants for coal and the move toward to sequestration of CO_2. It would not ban new coal plants outright. However, anyone planning to build a new conventional coal plant knows that their plant will not be grandfathered in to avoid future carbon costs. These new coal plants will have the financial burden of having to buy emission rights or pay a carbon emission fee. The emission trading market will find the value of the emissions, and it will have less to do with the "cost" of the carbon pollution and more to do with the cost of mechanisms to reduce carbon emissions. That could be a combination of things such as plant efficiency improvements, using lower carbon content fuel, or sequestering the CO_2. It could also be a combination of energy efficiency improvements in another sector of the economy, or paying tropical counties not to cut down their forests, etc. Whatever is the lower-cost approach to reducing carbon emissions plus the cost of administration of the complex cap-and-trade scheme, and somehow shielding it from political manipulation, it really is a carbon emission tax hidden in the form of a market of emission credits.

The other way to make carbon emissions expensive is a direct carbon emission charge via the revenue-neutral carbon dumping fee. Although you do have to calculate the amount of carbon a particular plant emits, it is a vastly simpler way to make carbon emissions cost something. But, how much should the carbon emission charge be?

One approach is for the carbon dumping fee to be priced to make a new conventional coal plants more expensive compared to the currently available commercial CC natural gas plant that emits half the CO_2. Based on current national average fuel prices [delivered cost of coal at $3.0/MBtu [180] and natural gas to electric utilities at $10/MBtu (2007) (EIA)], the carbon tax would have to be greater than $100/ton of carbon ($27/ton of CO_2). The energy cost of both plants would be 11.7 cents/kW·h with this fee compared to the no carbon fee cost of the natural gas plant cost of 10.3 cents/kW·h and the new coal plant cost of 7.5 cents/kW·h. Because the regional price of delivered coal varies from about $2/MBtu to $6.00/MBtu, half the new coal plants would still be cheaper than a CC natural gas plant at this nationally average breakeven carbon tax of $100/ton [181].

For the natural gas plant to be less than the new coal plant at the lowest regional delivered cost of coal, the carbon tax would have to be greater than about $130/ton carbon ($35/ton CO_2). Using this approach, the carbon tax would be sensitive to the delivered price of natural gas. If natural gas became more expensive, than the breakeven carbon tax would have to be incrementally above $130/ton for gas plants to be less expensive than a new coal plant. And pushing new plant construction to using natural gas would certainly put pressure on the price of natural gas.

When the sequestration sites are ready and the new coal plants are demonstrated, this approach can be used to compare a new conventional coal plant at 7.5 cents/kW·h to a new IGCC plant with carbon sequestration based on a projection of 10.1 cents/kW·h (coal national average delivered price of $3/ton). What would the carbon dumping fee need to be for the carbon sequestration plant to be the same cost? The revenue neutral carbon dumping fee would have to be the same as before, that is about $100/ton ($27/ton CO_2) if 90% of the CO_2 is captured. For a sequester coal plant to be less expensive than a new coal plant at the lowest regional coal price, the carbon tax would have to be more than $90/ton ($25/ton CO_2).

Using this approximate calculation as a guide, at a revenue neutral carbon dumping fee of $150/ton ($41/ton CO_2), the market would initially try to build CC gas plants that is a known technology with much commercial experience instead a new coal plant. This would reduce carbon emissions for each new natural gas plant by about half compared to the coal plant not built. As the sequestration sites were certified and several coal sequestration demonstration plants were built, coal sequestration plants would become a commercial option. This $150/ton carbon tax would be more than enough (greater than the breakeven $100/ton of carbon) for sequestration coal plants to be built compared a conventional coal plant or a CC gas plant at today's gas price. This would be especially true if the natural gas price rose above $10/MBtu. A properly priced carbon tax would drive new fossil plant construction first to efficient CC natural gas plants and then to sequestered coal plants once they are demonstrated.

This tax could be instituted over a 10-year period. For example, it could start in the first year at $75/ton ($20/ ton CO_2) and then be increased by 15% every 2 years until it is $150/ton ($41/ton CO_2) in 10 years. This predictable change in the rules of the carbon business would tend to mimic the strategy developed in Section 5.1. It would also let the individual cultural preferences of people function with these new market conditions without an overburden bureaucracy associated with the cap-and-trade approach. This vastly simpler mechanism deserves a full hearing. If presented as a "revenue-neutral carbon dumping fee" (not a tax) and is directly connected to reductions in pay-roll and other wealth producing taxes, it should be acceptable to most Americans. The concept of shifting taxes so that we tax bad things more than good things should not be a hard sell. It almost appears that the main thing that cap and trade has going for is it is that you do not call it a tax, although it essentially is a tax whose price is variable and determined by emission trading. The list of cap-and-trade negative features is long, and we should pause and ponder a long time before deciding to use this policy mechanism.

The other important difference between the carbon fee and cap-and-trade is that the tax will start having a significant impact on the day this provision is implemented. All coal-related financial decisions will take the $130/ton carbon fee into consideration, although it is introduced over 10 years. Coal decisions for the most part are long-term decisions because the power plants last 60 years or more. Cap-and-trade will have a lower cost of carbon initially, and the initial impact on coal plant decisions will be to increase the financial risk of a relatively unknown amount of the emission fee that will increase over decades in an unknown amount. Coal plants will still get built, but the financial costs will reflect this increased risk. It is difficult to estimate the rate that new coal plants will transition to other lower carbon choices. The early actions under cap-and-trade will be to pick the low hanging fruit, which will be to pay someone in South American not to chop down trees and someone in China to build a wind farm instead of a new coal power plant. These are all good things if they are "real"—how do you tell if the "plan" to cut down the trees or build a particular coal plant was real. However, the United States will build 40 instead of 50 new coal plants that will be pumping out CO_2 for 60 years with a cap-and-trade scheme. With the carbon fee implemented next year, the United States may not build any new traditional coal plants.

This study shows that the combination of energy efficiency and renewables is a cost-effective solution. The use of a revenue-neutral carbon fee addresses the issue of the currently tilted energy playing field and tends to level the field. It also addresses the need to make this transition sooner rather than later. As we move further away from carbon via these technical and policy recommendations, the impact of the extra cost of the carbon will abate as cost-competitive options are put in place.

CHAPTER 7

Call to Arms

Our species is faced with its most significant challenge ever. There have been historical examples of changes in the environment that have led societies to become extinct or dispersed due to the natural variability in the planet's climate. This is the first time that climate change has been activated by our collective actions. As the dominant animal species, we are dominating not only land use but most of the flora and fauna on the planet. Without realizing it until recently, this ascension to planetary dominance appeared like taking steps into a future that seemed to hold promise for most. These steps were marked by an improving standard of living for more and more people based on technology, better self-organization, and an increasing global market juggernaut. A gradual democratization of political institutions has occurred after millennia of tribal conflicts. Yes, with all this success, there are a number of issues needing attention, but these are problems that we are familiar with and are still facing with some hope of eventual success. This new issue of human-driven global climate change is so different in scope and complexity that it has taken a number of decades for us to hear what thousands of scientists reaching consensus decisions have been telling us with increasing clarity. The ball is now in our court. We, the people of the planet, have a clear challenge: significantly and quickly reduce carbon and other greenhouse gas emissions or catapult the planet's climate into a new climate regime fraught with difficulties and massive uncertainty.

This book does offer some hope in the face of overwhelming difficulties. Technical options and personal attitude change solutions are offered that appear to provide the emission reduction needed. These options appear to be cost-effective. It is not enough to just sit and wait for these carbon-free options to gradually enter wide usage. We need to mount a campaign to level the currently tilted playing field of energy options. More importantly, we need to be inclusive of the very different "energy tribes" that make up who we are, and take a path that keeps us all in the game. Many policies can work on paper, but they must work in the real world of multiculture value systems. A host of policy recommendations are made, and a serious attempt is made to check to see if they respect the different values contained in the body politic. If we maintain this balanced approach to policy and stay the course for the rest of this century, we have a pretty good chance of avoiding the worst of the possible damages that we are currently inflicting on the only home we have. Although this book focused on between now and 2050, what are the possibilities for a world operating on renewable energy sources? An earlier study looked at just this possibility and found that the global resources are more than adequate [182]. This global potential would encourage us to go beyond the 2050 goals to limit carbon dioxide—a world beyond the fossil era.

References

[1]   Deffeyes, K., *Beyond Oil: The View from Hubbert's Peak*, Hill and Wang, New York, 2005.

[2]   Hatfield, C., Oil back on the global agenda, *Nature* 387(1997):121.

[3]   Kerr, R.A., The next oil crisis looms large—and perhaps close. *Science* 281(1998):1128–1131. http://dx.doi.org/10.1126/science.281.5380.1128

[4]   Campbell, C., and J. Laherrere, The end of cheap oil. *Sci. Am.* 278(1998):78–83.

[5]   Campbell, C.J., *Oil Crisis*, Multi-Science Publishing, Essex, UK, 2005.

[6]   O'Neil, P., in R. Suskind, *The Price of Loyalty*, Simon & Schuster, New York, 2004.

[7]   Roberts, P., *The End of Oil: On the Edge of a Perilous New World*, Houghton Mifflin, Boston, New York, 2004.

[8]   Blimes, L., and Stiglitz, J., *The Economic Costs of the Iraq War*, ASSA Meetings, Boston, January 2006.

[9]   Verleger, P., *A Collaborative Policy to Neutralize the Economic Impact*, Policy Paper, Cambridge, MA, June 10, 2003.

[10]   Deffeyes, K., *The Second Great Depression*, Accessed on February 6, 2008. Available at www.princeton.edu/hubbert.

[11]   Gore, A., *An Inconvenient Truth: The Crisis of Global Warming*, Rodale, 2006.

[12]   *Smoke, Mirrors and Hot Air*, Union of Concerned Scientists, Cambridge, MA, January 2007.

[13]   Lindzen, R., Climate of fear, *Wall Street Journal*, Wednesday, April 12, 2006. Available at http://www.opinionjournal.com/extra/?id=110008220.

[14]   *Climate Change Science*, National Academy of Sciences, Washington, DC, June 6, 2001.

[15]   Gettleman, J., World Briefing: Africa: Kenya: Climate Deligates Agree to Review Kyoto Pact by 2008, *New York Times*, November 18, 2006, from the archives.

[16]   Hansen, J., *Testimony in the case of the Association of International Automobile Manufactures et al. v. the Secretary of Vermont*, August 14, 2006.

[17]   May, R., *Planet Earth BBC Special*, Oxford University, 5 March 2006.

[18]   Emanuel, K., Increasing destructiveness of tropical cyclones over the past 30 years, *Nature*, Vol. 436/4, August 2005.

[19]   Barnett, T. P., Pierce, D., W., AchutaRao, K., Gleckler, P., Santer, B., Gregory, J., and Washington, W., 2005: Penetration of Human-Induced Warming into the World's Oceans Science. Published online 2 June 2005, doi: 10.1126/science.1112418.

[20]   Sabine, C., The Oceanic Sink for Anthropogenic CO_2, *Science*, July 16, 2005, pp. 367–371.

[21]   Segre, G., A Matter of Degrees: What Temperature Reveals about the Past and Future of Our Species, Planet, and Universe, Viking, 2002.

[22]   Revkin, A., Arctic Ice Shelf Broke Off Canadian Island, *New York Times*, December 30, 2006, from archives.

[23]   Rignot, E., and Kanagaratnam, P., Changes in the velocity structure of the Greenland Ice Sheet, *Science* 311(5763) (February 17, 2006):986–990, doi:10.1126/science.1121381. http://dx.doi.org/10.1126/science.1121381

[24]   Schwartz, P., and Randall, D., *An Abrupt Climate Change Scenario and Its Implications for United States National Security*, Global Business Network, San Francisco, CA, February 2004.

[25]   Energy Information Agency, Department of Energy. Available at http://www.eia.doe.gov/oiaf/1605/ggccebro/chapter1.html.

[26]   ClimateChangeCorp.com report on James Hansen on the IPCC forecast August 16, 2007. Available at http://www.climatechangecorp.com/content_print.asp?ContentID=4898.

[27]   Scientific Expert Group Report on Climate Change and Sustainable Development, *Confronting Climate Change: Avoiding the Unmanageable and Managing the Unavoidable*, prepared for the 15th Session of the Commission on Sustainable Development, February 25–March 2, 2007.

[28]   Kutscher, C., ed., *Tackling Climate Change in the U.S.*, American Society of Solar Energy, Boulder, CO, January 2007.

[29]   Swisher, J.N., Potential carbon emissions reductions from overall energy efficiency by 2030, *Tackling Climate Change*, American Society of Solar Energy, Boulder, CO, January 2007.

[30]   *Emissions of Greenhouse Gases in the United States 2004*, Energy Information Administration, Office of Integrated Analysis and Forecasting, U.S. DOE, 2005

[31]   Hastings, M., The Gas Miser, *Newsweek*, September 20, 2004.

[32]   *Automotive Fuel Economy Program: Annual Update*, DOT HS 809 512, Calendar Year 2003, DOT, Washington, DC, November 2004.

[33]   Lilienthal, P., and Brown, H., Potential carbon emissions reductions from plug-in hybrid electric vehicles by 2030, in Kutscher, C., ed., *Tackling Climate Change in the U.S.*, American Society of Solar Energy, January 2008.

[34]   Quong, S., *The UCS Vanguard: A Vehicle to Reduce Global Warming Emissions Using Today's Technologies and Fuels*, Union of Concerned Scientists, Cambridge, MA, March 2007.

[35] Langer, G., *Poll: Traffic in the United States: A Look Under the Hood of a Nation on Wheels*, ABC News, February 13, 2005.

[36] Grahm, R., Comparing the Benefits and Impacts of Hybrid Electric Vehicle Options, EPRI, July 2001.

[37] Brown, M.A., Levine, M.D., Short, W., and Koomey, J.G., Scenarios for a clean energy future, *Energy Policy* 29(14) (2001):1179–1196.

[38] Tellus Institute. *1998 Policies and Measures to Reduce CO₂ Emissions in the United States: An Analysis of Options from 2005 and 2010*, Tellus Institute, Boston, MA, August 2005.

[39] Brown, M., Stovall, T., and Hughes, P., Potential carbon emissions reductions in the buildings sector by 2030, in Kutscher, C., ed., *Tackling Climate Change*, American Society of Solar Energy, Boulder, CO, January 2007, pp. 51–68.

[40] *Structural and Occupancy Characteristics of Housing: 2000*, U.S. Census Bureau, 2000.

[41] *No Reason to Wait: The Benefits of Greenhouse Gas Reduction in San Paulo and California*, Hewlett Foundation, Menlo Park, CA, December 2005.

[42] Brown, M., Southworth, F., and Stovall, T., *Towards a Climate-Friendly Built Environment*, Pew Center on Global Climate Change, Arlington, VA, 2005. Referenced May 2, 2006. Available at http://www.pewclimate.org/global-warming-in-depth/all_reports/buildings/index.cfm.

[43] *Scenarios for a Clean Energy Future*, ORNL/CON-476, Interlaboratory Working Group, Oak Ridge National Laboratory, Oak Ridge, TN, 2000.

[44] Lovins, A., et al., *Winning the Oil Endgame*, Rocky Mountain Institute, Snowmass, CO, 2004.

[45] Wisner, R., Bolinger, M., Annual Report on U. S. Wind Power Installation, Cost, and Performance Trends:, U. S. DOE, 2007.

[46] *Wind Turbine Interactions with Birds and Bats: A Summary of Research Results and Remaining Questions, Fact Sheet*, 2nd ed., NWCC, Washington, DC, November 2004.

[47] Klem, D., Jr., Collisions between birds and windows: mortality and prevention, *J. Field Ornithol.*, 61(1):120–128.

[48] Manville, A., U.S. Fish and Wildlife Service, briefing statement, April 7, 2000.

[49] Williams, R.H., and Weinberg, C.J., Energy from the sun, *Sci. Am.*, p. 98, September 1990.

[50] Milligan, M., Potential carbon emissions reductions from wind by 2030, in Kutscher, C., ed., *Tracking Climate Change*, American Society of Solar Energy, Boulder, CO, February 2007, pp. 101–112.

[51] Short, W.W., Blair, N., Heimiller, D., and Singh, V., *Modeling the Long-Term Market Penetration of Wind in the United States*, NREL/CP-620-34469, NREL, Golden, CO, 2003. Available at http://www.nrel.gov/docs/fy03osti/34469.pdf.

[52] *Multi-Year Program Plan 2007–2011, Solar Energy Technologies Program*, U.S. DOE, Washington, DC. Available at http://www1.eere.energy.gov/solar/pdfs/set_myp_2007-2011_proof_2.pdf.

[53] Caputo, R., et al., *Promise of Centralized Renewable Energy in the San Diego Region*, Report to SANDAG Resources Committee, April 3, 2006. Available at www.sdres.org.

[54] Mehos, M.S., NREL, and Kearney, D.W., Potential carbon emissions reductions from concentrating solar power by 2030, in Kutscher, C., ed., *Tackling Climate Change*, American Society of Solar Energy, Boulder, CO, February, 2007, pp. 79–90.

[55] Blair, N.N., Short, W., Mehos, M., and Heimiller, D., Concentrating Solar Deployment Systems (CSDS)—a new model for estimating U.S. concentrating solar power market potential, presented at the American Society of Solar Energy SOLAR 2006 conference, July 8–13, 2006.

[56] NREL Best Research, *Cell Efficiencies Graph Showing How Efficiency Has Increased from 1975 to 2004*, Sept 2004.

[57] Maycock, D., and Wakefield, G.F., *Business Analysis of Solar Photovoltaic Conversion*, The Conference Record of the 11th IEEE Photovoltaic Specialists Conference, IEEE, New York, 1975, p. 252.

[58] Harmon, C., *Experience Curves for Photovoltaic Technology, International Institute of Applied Systems Analysis*, IR-00-014, March 2000.

[59] Bradford, T., *Solar Revolution*, MIT Press, Cambridge, MA, 2006.

[60] Thomas, M., Post, H., and DeBlasio, R., Photovoltaic systems: an end-of-millennium review, in *Progress in Photovoltaics: Research and Applications*, vol. 7, John Wiley & Sons, Hoboken, NJ, 1999, pp. 1–19.

[61] Pernick, R., and Wilder, C., *Utility Solar Assessment (USA)*, Study Clean Edge Inc., Portland, OR, June 2008.

[62] Chaudhari, M., L. Frantzis, and T. Hoff, *PV Grid Connected Market Potential Under a Cost Breakthrough Scenario*, The Energy Foundation, San Francisco, CA, September 2004.

[63] Denholm, P., and Margolis, R., *Very Large-Scale Deployment of Grid-Connected Solar Photovoltaics in the United States: Challenges and Opportunities*, presented at the ASES SOLAR 2006 Conference, July 8–13, 2006.

[64] Zweibel, K., *The Terawatt Challenge for Thin-Film PV*, NREL/TP-520-38350, NREL, Golden, CO, 2005.

[65] Zweibel, K., Mason, J., and Fthenakis, V., A solar grand plan, *Sci. Am.*, January 2008.

[66] *TMY2 Users' Manual*, NREL, Golden, Colorado

[67] Swisher, J., *Tackling Climate Change in the U.S.: The Potential Contribution from Energy Efficiency*, presented at the ASES SOLAR 2006 National Conference, July 8–13, 2006.

[68] Denholm, P., Margolis, R., and Zweibel, K., Potential carbon emissions reductions from solar photovoltaics by 2030, in Kutscher, C., ed., *Tackling Climate Change in the U.S.*, American Society of Solar Energy, Boulder, CO, January 2007, pp. 91–100.

[69] Tester, J., et al., *The Future of Geothermal Energy—Energy Recovery from Enhanced/Engineered Geothermal Systems (EGS)—Assessment of Impact for the United States by 2050*, MIT, Cambridge, MA, 2006.

[70] Vorum, M., and Tester, J., Potential carbon emissions reductions from geothermal power by 2030, in Kutscher, C., ed., *Tackling Climate Change in the U.S.*, American Society of Solar Energy, Boulder, CO, February, 2007, pp. 145–162.

[71] Edwards, R., Larivè, F., et al., *Well-to-Wheels Report, Version 2a*, European Commission Joint Research Centre, CONCAWE and EUCAR 88, Ispra, Italy, 2005.

[72] Milbrandt, A., *A Geographic Perspective on the Current Biomass Resource Availability in the United States*, NREL/TP-560-39181, NREL, Golden, CO, 2005.

[73] Overend, R.P., and Milbrand, A., Potential Carbon Emissions Reductions from Biomass in 2030, in Kutscher, C., ed., *Tracking Climate Change in the U.S.*, American Society of Solar Energy, Boulder, CO, January 2007, pp. 113–130.

[74] Perlack, R.R., Wright, L., et al., *Biomass as Feedstock for a Bioenergy and Bioproducts Industry: The Technical Feasibility of a Billion-Ton Annual Supply*, ORNL/TM-2005/66, Oak Ridge National Laboratory, Oak Ridge, TN, 2005.

[75] Biomass Task Force Report, *Supply Addendum, Clean and Diversified Energy Initiative 51*, Western Governors' Association, Denver, CO, 2006.

[76] Nilsson, D., SHAM—a simulation model for designing straw fuel delivery systems, Part 1: Model description, *Biomass and Bioenergy* 1999;16(1):25–38.

[77] Overend, R., The average haul distance and transportation work factors for biomass delivered to a central plant, *Biomass* 2(1982) 75–79.

[78] Statistical Abstract of the United States (1997).

[79] Short, W., Packey, D., et al., *A Manual for the Economic Evaluation of Energy Efficiency and Renewable Energy Technologies*, NREL, Golden, CO, 1995.

[80] *Quantitative Work Group Guidance to Task Forces of the WGA's CDEAC, Version 1.6*, July 12, 2005.

[81] Liebig, M., Johnson, H., et al., Soil carbon under switchgrass stands and cultivated cropland, *Biomass Bioenergy* 28(4) (2005)347.

[82] Davidson, E., Trumbore, E., et al., Soil warming and organic carbon content, *Nature* 408(6814) (2000):789–790.

[83] Davidson, E., and I. Janssens. Temperature sensitivity of soil carbon decomposition and feedbacks to climate change, *Nature* 440(7081) (2006):165.

[84] David Pimentel et al., *Bioscience*, 55:7.

[85] R. Lal, J.M. Kimble, R.F. Follett, and C.V. Cole, *The Potential of U.S. Cropland to Sequester Carbon and Mitigate the Greenhouse Effect*, Ann Arbor Sci. Publisher, Chelsea, MI, 1998.

[86] McCarl, B., and U. Schneider, The cost of greenhouse gas mitigation in U.S. agriculture and forestry, *Science*, 294(5551) (2001):2481–2482.

[87] Antle, J.M., and B.A. McCarl, The economics of carbon sequestration in agricultural soils, in T. Tietenberg and H. Folmer, ed., *The International Yearbook of Environmental and Resource Economics 2002/2003: A Survey of Current Issues*, Edward Elgar, Northampton, MA, 2002, pp. 278–310.

[88] Manley, J., G.C. van Kooten, K. Moeltner, and D.W. Johnson, Creating carbon offsets in agriculture through no-till cultivation: a meta analysis of costs and carbon benefits, *Climate Change*, 68(1) (2005):41–65.

[89] *Annual Energy Outlook 2007: Projections to 2030*, EIA, U.S. DOE, Washington, DC, 2006. Available at http://www.eia.doe.gov/oiaf/aeo/pdf/trend_4.pdf.

[90] Kansas Corn Growers Association. Available at http://www.ksgrains.com/ethanol/useth .html.

[91] Bantz, S., *Biofuels: An Important Part of a Low Carbon Diet*, Union of Concerned Scientists, Cambridge, MA, November 2007.

[92] K. Collins. *The Role of Biofuels and Other Factors in Increasing Farm and Food Prices*, Kraft Foods Global, Winnetka, IL, June 19, 2008.

[93] Sheehan, J., et al., Energy and environmental aspects of using corn stover for fuel ethanol, *J. Ind. Ecol.*, 7(2005):3–4.

[94] Wang, M., et al., Well-to-Wheels Analysis of Advanced Fuel/Vehicle Systems—A North American Study of Energy Use, Greenhouse Gas Emissions, and Criteria Pollutant Emissions, Argonne National Laboratory, Argonne, IL, 2005.

[95] Sheehan, J., Potential carbon emissions reductions from biofuels by 2030, in Kutscher, C., ed., *Tackling Climate Change in the U.S.*, American Society of Solar Energy, Boulder, CO, January 2007, pp. 131–144.

[96] Aden, A., Ruth, M., et al., *Lignocellulosic Biomass to Ethanol Process Design and Economics Utilizing Co-Current Dilute Acid Prehydrolysis and Enzymatic hydrolysis for Corn Stover*, NREL/TP-510-32438, NREL, Golden, CO, 2002.

[97] Lynd, L., et al., *The Role of Biomass in America's Energy Future*, prepared for the NREL, 2005.

[98] Rudervall, R., Chairpentier, J.P., Sharm, R., *High Voltage Direct Current (HVDC) Transmission Systems Technology Review Paper*, World Bank.

[99] *Frontier Line Analysis of Transmission Links and Costs to be Used by the Economic Analysis Subcommittee, Western Regional Transmission Expansion Partnership Transmission Subcommittee,* Final Report, Dec 1, 2006.

[100] Kutscher, C., Overview and summary, in Kutscher, C., ed., *Tackling Climate Change in the U.S.,* American Society of Solar Energy, Boulder, CO, January 2007, pp. 5–38.

[101] Kutscher, C., Tackling climate change: can we afford it? *Solar Today,* May 8.

[102] Bezdek, R., *Renewable Energy and Energy Efficiency: Economic Drivers for the 21st Century,* Management Information Services Inc., published by American Society of Solar Energy, Boulder, Colorado, January 2007.

[103] Caputo, R., *An Initial Comparative Assessment of Orbital and Terrestrial Central Power Systems, Final Report,* Jet Propulsion Laboratory, Pasadena, CA, 900-780, March 1977.

[104] Marchetti, C., and Nakicenovic, N., *The Dynamics of Energy Systems and the Logistic Substitution Model,* RR-79-13, IIASA, Laxenburg, Austria, December 1979.

[105] Goodell, J., *Big Coal, A Mariner Book,* Houghton Mifflin, Boston and New York, 2007.

[106] *Coal Fatalities for 1900 Through 2004,* U.S. Department of Labor, Mine Safety and Health Administration. Available at www.msha.gov/stats/centurystats/coalstats.htm.

[107] National Institute for Occupational Safety and Health at www.courierjournal.com/dust/illo_lungdeaths.html.

[108] American Lung Association.

[109] *2000 Resource Assessment of Selected Coal Beds and Zones in the Northern and Central Appalachian Basin Coal Regions,* U.S. Geologic Survey, Washington, DC, updated May 2003.

[110] Shnayerson, M., The rape of Appalachia, *Vanity Fair,* May 2006, from the archives.

[111] *Coal-Mine-Drainage Projects in Pennsylvania,* U.S. Geological Survey, Washington, DC, updated January 2007.

[112] *EPA to Regulate Mercury and Other Air Toxics Emissions from Coal- and Oil-Fired Power Plants,* U.S. Environmental Protection Agency, Washington, DC, December 14, 2000.

[113] Agency for Toxic Substances and Disease Registry, *ToxFAQs for Mercury,* U. S. Department of Health and Human Services, Washington, DC, April 1999.

[114] *Mercury, Fish Oils, and Risk of Acute Coronary Events and Cardiovascular Disease, Coronary Heart Disease, and All-Cause Mortality in Men in Eastern Finland,* American Heart Association, Dallas, TX, November 11, 2004.

[115] *Briefing: How Coal Works,* Union of Concerned Scientists. Cambridge, MA, April 2007. Available at http://www.ucsusa.org/clean_energy/coalvswind/brief_coal.html.

[116] *Inventory of U.S. Greenhouse Gas Emissions and Sinks: 1990–2004,* U.S. Environmental Protection Agency, Washington, DC, April 2006.

[117]   *New York Times*, 1970.

[118]   EIA Annual Energy Outlook 2008. Available at http://www.eia.doe.gov/oiaf/aeo/aeoref_tab.html.

[119]   *Living on Earth*, January 4, 2008. Available at http://www.loe.org/shows/segments.htm?programID=08-P13-00001&segmentID=5.

[120]   Jordall, K., Anheden, M., Yan, J., and Strömberg, L., *Oxygen Fuel Combustion for Coal-Fired Power Generation with CO_2 Capture—Opportunities and Challenges*, Vattenfall Utveckling AB, Stockholm, Sweden.

[121]   Dooley, J., *On the Potential Large-Scale Commercial Deployment of Carbon Dioxide Capture and Storage Technologies: Findings from Phase 2 of the Global Energy Technology Strategy Project PNNL-SA-53472*, Joint Global Change Research Initiative, Battelle, February 2007.

[122]   New power plant aims to help coal cleanup, *Sci. Am.*, December 2007.

[123]   Up in Smoke, *The Economists*, January 31, 2008.

[124]   Lee, H., and Verma, S.K., *Coal or Gas: The Cost of Cleaner Power in the Mid-West*, E-2000-08, Harvard University John F. Kennedy School of Government, Environmental and Natural Resources Program Belfer Center for Science and International Affairs, August 2000.

[125]   Lee, L., Banks to weigh CO_2 emissions in power lending, Reuters, New York, February 4, 2008.

[126]   Kansas Power Plant Permit Rejected on Grounds of Carbon Emissions, San Francisco Sentinel, 19 October 2007.

[127]   *The Future of Nuclear Energy: An Interdisciplinary Study*, MIT, Cambridge, MA, 2003.

[128]   Time/CNN, June 27, 1983.

[129]   Caputo, R., *An Initial Comparative Assessment of Orbital and Terrestrial Central Power Systems, Final Report*, Report 900-780, Jet Propulsion Laboratory, Pasadena, CA, March 1977.

[130]   Lyman, E.S., *Chernobyl-on-the-Hudson? The health and economic impacts of a terrorist attack at the Indian Point Nuclear Plant*, Riverkeeper, Tarrytown, NY, September 2004.

[131]   NRC, Status of Accident Sequence Precursor and SPAR Model Development Programs, SECY-02-0041, March 8, 2002.

[132]   *Severe accident mitigation alternatives (SAMA) analysis*, Guidance Document, NEI 05-01 (Rev. A), Nuclear Energy Institute, Washington, DC, November 2005.

[133]   Committee on Science and Technology for Countering Terrorism, *Making The Nation Safer: The Role of Science and Technology in Countering Terrorism*, National Academy of Sciences, Washington, DC, 2002. Available at http://www.nap.edu/books/0309084814/html.

[134]   Gronlund, L., Lochbaum, D., and Lyman, E., *Nuclear Power in a Warmer World*, Union of Concerned Scientists, Cambridge, MA, December 2007.

[135]   Caldicott, H., *Nuclear Power Is Not the Answer*, The New Press, New York, 2006.

[136]   National Research Council, Board on Radioactive Waste Management, *Safety and Security of Commercial Nuclear Fuel Storage*, executive summary, National Academies Press, Washington, DC, 2005, p. 6.

[137]   Beyea, J., Lyman, E., and von Hippel, F., Damages from a major release of ^{137}Cs into the atmosphere of the United States, *Sci. Global Security* 12(2004):125–136. http://dx.doi.org/10.1080/08929880490464775

[138]   Travis, R.J., Davis, R.E., Grove, E.J., and Azarm, M.A., *Brookhaven National Laboratory. A Safety and Regulatory Assessment of Generic BWR and PWR Permanently Shutdown Nuclear Power Plants*, NUREG/CR-6451, August 1997.

[139]   Westinghouse Electric Co. Available at http://www.ap600.westinghousenuclear.com/plant.htm.

[140]   Lewis, E.E., *Nuclear Power Reactor Safety*, Wiley, New York, 1977.

[141]   Craig, P., Blowing the whistle on Yucca Mountain, *Energy Bulletin*, June 5, 2004. Available at http://www.energybulletin.net/node/505.

[142]   Nevada State web site, *Chronology of Selected Yucca Mountain Emails*, http://www.state.nv.us/nucwaste, September 9, 2005.

[143]   *The Future of Nuclear Energy: An Interdisciplinary Study*, MIT, Cambridge, MA, 2003.

[144]   Presentation slides for staff briefing to executive director for operations, Status of NRC staff review of FENOC's Bulletin 2001–01 Response for Davis-Besse, National Research Council, Washington, DC, November 29, 2001.

[145]   NRC Inspection Report 50-346/2003-16, National Regulatory Commission, Washington, DC.

[146]   Williams, P.T., Yin, S., and Bass, B.R., *Probabilistic Structural Mechanics Analysis of the Degraded Davis-Besse RPV Head*, Oak Ridge National Laboratory, Oak Ridge, TN, September 2004.

[147]   Luers, A., Mastramdrea, M., Hayhoe K., Frumhoff, P., *How to Avoid Dangerous Climate Change*, Union of Concerned Scientists, Sept. 2007.

[148]   Keeter, S., Masci, D., Smith, G., *Science in America: Religious Belief and Public Attitudes*, Pew Research Center Publications, Washington, DC, December 18, 2007.

[149]   International Panel on Climate Change Report (2007).

[150]   Stern Review Report on the Economics of Climate Change. Available at http://www.hm-treasury.gov.uk/independent_reviews/stern_review_economics_climate_change/stern_review_Report.cfm.

[151]   Benedict, R., *Patterns of Culture*, A Mariner Book, Houghton Mifflin, Boston, New York, 1934.

[152]   Douglas, M., and Wildavsky, A., *Risk and Culture*, University of California Press, Berkeley, CA, 1983.

[153] Thompson, M., and Rayner, S., Cultural discourses, in Rayner., S., and Malone, E.L., eds., *Human Choice and Climate Change*, vol. 1, p. 333, Battelle Press, Columbus, OH, 1998.

[154] Verweig, M., and Thompson, M., eds., *Clumsy Solutions for a Complex World*, Palgrave Macmillan, Hampshire and New York, 2006.

[155] Caputo, R., Energy worlds in collision: is a rational energy policy possible for countries in Western Europe, *Futures*, Butterworth & Co., June 1984, from archives.

[156] *IASA Options*, Fall 2006.

[157] Verweij, M., Douglas, M., et al., The case of clumsiness, in Verweij, M., and Thompson, M., eds., *Clumsy Solutions for a Complex World*, Palgrave Macmillan, Hampshire and New York, 2006, pp. 1–30.

[158] Ney, S., and Thompson, M., Consulting the frogs: the normative implications of cultural theory, in Thompson, Grendstad, and Selle, eds., *Cultural Theory as Political Science*, Routledge, 1999, from archives.

[159] Banfield, E., *The Moral Basis of a Backward Society*, Free Press, New York, 1958.

[160] Douglas, M., Cultural bias, in *Occasional Paper No. 35*, Royal Anthropological Institute, London, 1978.

[161] Scott, J., *Seeing Like a State: How Certain Schemes to Improve the Human Condition Have Failed*, Yale University Press, New Haven, CT, 1999.

[162] Bate, R., The political economy of climate change science, in *Environmental Unit Briefing Paper No. 1*, Institute of Economic Affairs, London, 2001.

[163] Hendriks, F., and Zouridis, S., Cultural biases and the new public domain, in Thompson, Grendstad, and Selle, eds., *Cultural Theory as Political Science*, Routledge, 1999.

[164] Veweij, M., Is the Kyoto Protocol merely irrelevant, or positively harmful, for the efforts to curb climate change?, in Verweij, M., and Thompson, M., eds., *Clumsy Solutions for a Complex World*, Palgrave Macmillan, Hampshire and New York, 2006.

[165] Jonas, .M., Obesteiner, M., and Nillson, S., *How to Go from Today's Kyoto Protocol to a Post-Kyoto Future that Adheres to the Principles of Full Carbon Accounting and Global-Scale Verifications*, Interim Report IR-00-61, IIASA, Laxenburg, Austria, 2000.

[166] Greenhouse Gas Commission of the European Communities, *Green Paper on Greenhouse Gas Emissions Trading within the European Union*, European Commission, Brussels, May 2006.

[167] *UNEP/OECD/IEA Workshop on Baseline Methodologies, Possibilities for Standardized Baselines for JI and CDM: Chairman's Recommendations and Workshop Report*, OECD, Paris, July 2001.

[168] Cleetus, R., *We Need a Well-Designed Cap and trade Program to Fight Global Warming*, Union of Concerned Scientists, Cambridge, MA, October 2007.

[169]   Hayes, D., Climate Solutions: Charting a Bold Course, *Environment 360*. Available at www.e360.yale.edu/content/print.msp?id=2026.

[170]   Gore testimony to House of Representatives, March 21, 2007.

[171]   Opinion, *L.A. Times*, May 28, 2007.

[172]   Bailie, A., et al., *2003, Analysis of the Climate Stewardship Act*, Tellus Institute, Boston, MA. Avaiable at www.tellus.org/energy/publications/McCainLieberman2003.pdf.

[173]   Dooley, J., Ruci, P., and Luiten, E., *Energy RD&D in the Industrialized World: Retrenchment and Refocusing*, Interim Report PNNL-12661, Pacific Northwest National Laboratory, Richland, WA, 1998.

[174]   Watson, G., and Courtney, F., Nantucket Sound offshore wind stakeholder process, *Boston College Environmental Affairs*, 31(2)(2004).

[175]   Brewer, G.D., *Hydrogen Aircraft Technology*, NASA 19930049114, 1991.

[176]   Mims, N., and Hauenstein, H., *Feebates: A Legislative Option to Encourage Continuous Improvements to Automobile Efficiency*, Rocky Mountain Institute, February 2008. Available at http://www.rmi.org/images/PDFs/Transportation/Feebate_final.pdf.

[177]   Pullins, S., and Westerman, J., *San Diego Smart Grid Study*, The Energy Policy Initiatives Center, School of Law, University of San Diego, San Diego, CA, October 2006.

[178]   Anders, S., et al., *California's Solar Rights Act: A Review of the Statutes and Relevant Cases*, Energy Policy Initiatives Center, School of Law, University of San Diego, San Diego, CA, January 2007.

[179]   Pollan, M., Why bother? *New York Times Magazine*, April 20, 2008, from the archives.

[180]   Average Weekly Coal Commodity Spot Prices | U. S. Monthly Coal Production | 12-Month Eastern Coal Production Trends | Electric Power Sector Coal Stocks | Average Cost of Metallurgical Coal, EIA, 2008. Available at http://www.eia.doe.gov/cneaf/coal/page/coalnews/coalmar.html#spot.

[181]   Hong, B.D., and Slatick, E.R., *Carbon Dioxide Emission Factors for Coal*, EIA, 1994. Available at http://www.eia.doe.gov/cneaf/coal/quarterly/co2_article/co2.html.

[182]   Caputo, R., *Solar Energy for the Next 5 Billion Years*, Analysis PP-81-9, IIASA, Laxenburg, Austria, May 1081.

Bibliography

Alley, R., et al., *Abrupt Climate Change: Inevitable Surprises*, National Academies Press, Washington, DC, 2002.

Bent, R. D., et al., ed., *Energy: Science, Policy, and the Pursuit of Sustainability*, Island Press, Washington, DC, 2002.

Boyle, G., ed., *Renewable Energy: Power for a Sustainable Future*, Oxford University Press, 2004.

Boyle, G., Everett, B., and Ramage, J., *Energy Systems and Sustainability*, Oxford University Press, 2003.

Brown, L., *Plan B 2.0: Rescuing a Planet Under Stress and a Civilization in Trouble*, W. N. Norton and Company, 2006.

Casten, T., *Turning Off the Heat*, Prometheus Books, 1998.

Davis, D. H., *Energy Politics*, St. Martin's Press, 1993.

Deutch, J., and Moniz, E., *The Future of Nuclear Energy: An Interdisciplinary Study*, MIT Press, Cambridge, MA, 2003.

Geller, H., *Energy Revolution: Policies for a Sustainable Future*, Island Press, Washington, DC, 2003.

Huber, P. W., and Mills, M. P., *The Bottomless Well: The Twilight of Fuel, the Virtue of Waste, and Why We Will Never Run Out of Energy*, Basic Books, 2005.

Jaffe, A., et al., eds., *Strategic Energy Policy: Challenges for the 21st Century Independent Task Force Report*, Task Force Report, Council on Foreign Relations, Washington, DC.

Kolbert, E., *Field Notes for a Catastrophe, Man, Nature, and Climate Change*, Bloomsbury, 2006.

Komor, P., *Renewable Energy Policy*, Diebold Institute for Public Policy Studies, iUniverse, New York, 2004.

National Commission on Energy Policy, *Ending the Energy Stalemate*, NCEP, Washington, DC, 2004.

Smil, V., *Energy at the Crossroads: Global Perspectives and Uncertainties*, MIT Press, Cambridge, MA, 2003.

Stern, N., Stern Review Report on the Economics of Climate Change, Cambridge University Press, Cambridge, UK, 2006.

Stern, P., and Eastering, W., eds., *Making Climate Forecasts Matter*, National Academies Press, Washington, DC, 1999.

Index

Printed in the United States
by Baker & Taylor Publisher Services